Technics and Enaction

Also Available from Bloomsbury

The Philosophy of Imagination, ed. Galit Wellner , Geoffrey Dierckxsens, Marco Arienti

Bernard Stiegler, Bart Buseyne

Technics and Enaction

A Philosophy of Imagination

Émilien Dereclenne

BLOOMSBURY ACADEMIC
LONDON • NEW YORK • OXFORD • NEW DELHI • SYDNEY

BLOOMSBURY ACADEMIC
Bloomsbury Publishing Plc
50 Bedford Square, London, WC1B 3DP, UK
1385 Broadway, New York, NY 10018, USA
29 Earlsfort Terrace, Dublin 2, Ireland

BLOOMSBURY, BLOOMSBURY ACADEMIC and the Diana logo are trademarks of Bloomsbury Publishing Plc

First published in Great Britain 2025

Series design by Charlotte Daniels
Cover image: Agathe Herry, photographed by Sandrine Binoux

A catalogue record for this book is available from the British Library.

A catalog record for this book is available from the Library of Congress.

ISBN: HB: 978-1-3505-0759-3
ePDF: 978-1-3505-0760-9
eBook: 978-1-3505-0761-6

Typeset by Deanta Global Publishing Services, Chennai, India

Contents

Introduction

Enacting Imagination

In his famous *Discourse on the Origin and Basis of Inequality Among Men* (1755), Jean-Jacques Rousseau describes the so-called "state of nature" as a morally neutral and peaceful condition, where human beings relate to each other regardless of any political institution. In *Le Geste et la parole* (1964/1977, p. 19), French anthropologist André Leroi-Gourhan denounces this Rousseauist view as delusory. He dismisses what he takes to be a naïve and misleading vision, that of a man who would define himself in his actual characteristics, independently from what he is as a socio-technical being. As the expression goes, there is no "zero degree of man." The natural man, taken independently from the tools, instruments, interfaces, socio-material organizations and institutions, technologies, and information and communication systems that we design, develop, and use, is a misleading tale. A creature in a state of pure nature could not possibly possess all the current attributes of human beings. Leroi-Gourhan also emphasizes that, rather than an "effect," technical development is a "corollary" of humanization. Basically, an effect *y* is a change caused by a cause *x*, where *x* is temporally and ontologically prior to *y*. By contrast, a corollary is something *y* resulting from something else *x*, where the very genesis and existence of *x* at the same time depends on the genesis and existence of *y*. In Leroi-Gourhan's perspective, a perspective shared by other contemporary anthropologists (see, for instance, Eric Boeda 2021), humanization does not support technical development. Instead, both anthropogenesis and technogenesis constitute each other, in a dialectical process of psychological and socio-material co-individuation. The confusion between "effect" and "corollary" explains that we traditionally phrase in naïve terms a question I will thematize in the following pages, that of the relation between brain, psychology, and technical development. The question is not "What brain and what mind did it take for technical development to be possible?" but, rather, "What technical device is needed for such brain and psychological capacity to appear in the world?"

In this perspective, the study of human psychology cannot be separated from the study of the history of couplings between living beings and their socio-material, especially technical, environment. And as a matter of starting observation, in the field of contemporary cognitive science, no real thought about the relation between imagination and technics is to be found. Nowhere is the question of the technical constitutivity of imagination thematized. It was only recently that the question was touched upon in the manner of a theoretical project, by enactive philosophers of cognitive science (Malafouris 2013; Hutto and Myin 2017; Gallagher 2017). In the wake of it, proponents of 5E cognition (enactive, embodied, embedded, extended, ecological) approaches started taking the role of socio-material and technical engagement in imagination more seriously (Poulsgaard 2019; Van Dijk and Rietveld 2020; Koukouti and Malafouris 2020a, 2020b). This book clearly has a vocation to pursue this reflection.

What I intend to show is that imagination is enacted through technical engagement—nothing of a pre-constituted faculty which we magically possess from the start. This claim raises important philosophical questions, which I shall address successively in the following chapters. In Chapter 1, as a state of the art, I show that contemporary conceptions of imagination and creativity in the field of cognitive science, mainly internalist and representationalist, seriously underestimate the role of technical and socio-material engagement in imaginative and creative processes. Guilty of the kind of dualism between mind and technics which Leroi-Gourhan invites us to get rid of, internalist and representationalist conceptions of imagination and creativity classically understand technical development and engagement as a means for the exteriorization of imaginative and creative achievements that remain internal to the only representational mind-brain, prior to concrete embodied engagement with the technical environment. In this theoretical context, technics (tools, instruments, interfaces, physical organizations, institutions, technologies) does not constitute, or participate in, in any way, imaginative and creative processes.

In Chapter 2, I present the epistemological (Section 1) and ontological (Section 2) principles by which to think about the relationship between imagination and technics and exceed the limitations of internalism and representationalism. In Section 1, I diagnose with pragmatism, especially in the light of John Dewey and Nelson Goodman's epistemological criticisms, the lack of consideration for technics in the contemporary cognitive science of imagination and creativity, as being the result of a philosophical fallacy, the so-called "mereological fallacy." To put it quickly here, this mereological fallacy consists in attributing to the mind-brain the only explanatory role, ignoring the essential and constitutive role of external factors and of practical engagement. I explain how this fallacy works in the case of imagination and argue that there is another way to explain imaginative processes than postulating the existence of mental representations and reducing imaginative processes to purely internal and brain processes.

In Section 2, I show that this epistemological criticism leads to the promotion of a pragmatist ontology of individual-world transactions, which I nourish in Part Two and Three of this book, with Gilbert Simondon's ontology of relations, the enactive ontology of individual-world couplings and the ecological ontology of affordances. Representationalism and internalism rest on a dualistic ontology, grounded on conceptual oppositions such as internal-external, mind-world, and subjective-objective. These arbitrary oppositions prevent us from thinking of imagination and creativity as technically constituted. In contrast to this, I adopt an ecological view on cognition and imagination. I introduce the idea that, instead of locating imagination and creativity in a brain or a mind filled with mental representations, with "mental images" as we usually call them, we had better understand cognition, in particular imagination, in terms of a whole practical and dynamical system, an ecological system spanning over the brain, the body and the socio-material environment. The idea is that extending cognition into the world and pluralizing the functional resources of cognition beyond the brain, beyond purely natural, neurophysiological processes, enables us to unburden this very brain of expensive representational tasks, whose effectivity could not be explained independently from technical engagement and the manipulation of external representations. In the wake of Goodman and Wittgenstein's perspectives, I argue that "mental images," as we use to define them in terms of representational states located in the brain, simply do not exist. Instead of talking about mental images in the brain, I define imaginative experiences in terms of a continuous dynamical interaction between the brain, the body, and the temporally extended and technically constituted socio-material environment.

Part Two nourishes this view with different but converging philosophical insights. This second part of the book is more circular than incremental in its progression. I bet on a crossed reading of the analytic and the continental philosophical traditions to propose a comprehensive theory of imagination and creativity as technically constituted. I progressively define what a relation of technical "constitution" is, in contrast with a relation of "realization" and a relation of "reduction." I show that in the context of these nondualistic and nonreductionist ontologies, classical oppositions like internal/external, subjective/objective, mind/world, simply become obsolete. In Chapters 3 to 5, I bring to light the strong theoretical and conceptual choices that allow material anthropologist Tim Ingold, French philosophers of technics Gilbert Simondon, Jacques Derrida, and Bernard Stiegler, and cognitive archeologist Lambros Malafouris, to apprehend, beyond those classical oppositions, the constitutive relation between imagination and technics.

Chapter 3 gives me the opportunity to put Gilbert Simondon's philosophical perspective at work in the field of 5E cognition approaches. In Section 1, I present Simondon's conception of individuation. In Section 2, I explain how Simondon can help enactivists to think of the relation between life, imagination, and technics. Simondon appeared to me as a thinker to start with for at least three reasons. First, he thematizes the articulation between imagination and technics, as a key to understanding psychic and collective individuation. Second, he inspired Tim Ingold and Bernard Stiegler in some of their most important arguments and conceptual developments. And finally, he is one of the most enactive-friendly authors of the twentieth-century French philosophy of life, subjectivity, and technics. Like enactivists, Simondon thematizes the insertion of subjectivity in life. And he offers stimulating perspectives to help enactivists think the articulation, not only, as classical enaction does, between life, cognition, and the lived body but more extensively between life, cognition, and the sociocultural environment. For these reasons, Simondon has already been brought into enaction (Stewart, Lenay, and Havelange 2003; Stewart 2010; Di Paolo, 2018, 2021; Di Paolo, Cuffari, and De Jaegher 2018; Poulsgaard 2019; Dereclenne 2020; Garcia and Arandia 2022), and I pursue the same line of thought by focusing on imagination.

With Simondon, I thematize two things. First, the idea that imagination, instead of a function by means of which we escape reality, is a function of realization, whereby we constantly engage with the world as organisms. Simondon's theory of imagination takes place in the core of a wider reflection on life and on the so-called "transductive," instead of "adaptive," relation between living beings and their environment. Simondon's originality lies in that he explains this transductive relation in terms of "imagination" and "image cycle." In Simondon, images are pre-reflexive motor structures and spontaneities, patterns of action-perception through which organisms affectively engage with, and grasp the meaning of, their perceptive world.

Defining imagination as a function of realization, instead of as a function of irrealization, is a strong philosophical gesture. But it is not a new one. Let's think of the Renaissance thinkers, in particular of Giordano Bruno, in books II and III of his *De imaginum compositione* (1591), who understands imaginative phenomena in terms of interiorization of technical acts of notation and classification (see Chapter 7). Let us think of twelfth-century Arab philosopher Ibn al Arabi and his notion of "imaginal" as an organ of subjective perception whereby individual consciousness opens to the world, such as it was thematized by Henry Corbin and, at the same time (1940–70; see Chittick 1994), by an important reference of Simondon, namely Carl Gustav Jung; let's think of Kant's definition of imaginative schematism as a transcendental condition of subjective experience (see Chapters 5 and 7), of Henri Bergson's dynamical imagination (see Chapter 7), of French philosopher of science Gaston Bachelard's awake dreamers (see Chapters 4 and 7). As much philosophical theorizations, among others, of imagination as the very source and dynamical foundation of experience. What these approaches to imagination suggest is that imagination is more than what contemporary cognitive scientists generally define in terms of an ability we have to possess, manipulate, and produce mental images of things that are not present in the field of perception. According to this widely shared view, imagination unleashes us from reality. Instead of perceiving real things, we mentally visualize fictional and fanciful ones. Imagination disengages us from the concrete experience of things. Such a reduction, however, of imagination to its fictional function, to its function of unreality, simply is a nonsense with regard to the history of its notion.

In Simondon's perspective, the only way to understand the constitutive relation between imagination and technics lies in loosening the grip of a narrow concept of imagination that reduces it to the consideration of counterfactual truth and fiction. I will give particular attention to Simondon's notion of "practical schematism." Practical schematism refers to how technogenesis and image genesis constantly relate to each other, as the two faces of a same coin. Put simply, for Simondon imagining consists in engaging with external genetic processes that are natural and technical, always material, and independently from which imagination and creativity could not be explained.

In Chapter 4, I pursue this reflection on imagination and creativity with material anthropologist Tim Ingold. Ingold shares with Simondon the idea that imagination is a mode of engagement with the world. He rejects the hylomorphic schema, according to which imagination takes place before and independently from material engagement. In Ingold's perspective, this goes along with a radical rejection of internalism and representationalism. According to Ingold, imagination cannot be understood separately from the external and technical (especially instrumental) mediations (tools, instruments) by means of which the living being imaginatively and creatively engages with its material world. Creativity has nothing to do with the *a posteriori* materialization of a purely mental and representational design. Rather, it must be understood as the outcome of a manipulative relation between the agent and its material and technical milieu. I illustrate this point in Section 2, through the analysis of an example dear to Ingold's heart and to mine, that of cello bowing.

In the wake of it, in Chapter 5, I argue that imagination, contrary to a timeless and world-separated faculty to produce novelty and mental images, refers to a technically constituted disposition to produce material images like drawings, paintings, pictures, musical performances, architectural designs, and so on—a disposition that renews itself through the history of the individual's material and technical engagement. In this perspective, directly inspired by Bernard Stiegler's organology and theory of imaginative schematism (Section 1), and by Lambros Malafouris' Material Engagement Theory (MET, Section 2), the study of imagination is inseparable from the study of the technical devices whereby material images appear in the world. This amounts to emphasizing the technical relativity of imagination and its irreducibility to purely internal and representational processes, isolated in time and from the technical environment.

This demands renouncing the idea that imagination is a faculty. The term "faculty" is problematic because it suggests the existence in the soul of several mysterious entities, each of which being distinct from the others (memory, imagination, will, understanding, and so on), each endowed with a proper power. However, defining along with Simondon and enaction, imagination as a mode of bodily and technical engagement with the world, amounts to understanding it as a multifaceted behavior which borders with other behaviors are not always as clear as one might think (perception, action, socio-technical engagement). Furthermore, articulating imagination to technical engagement amounts to acknowledging the existence of a diversity of imaginative modes, dependent on practical, technically, and socially constituted situations that enable specific kinds of imaginative phenomena. In this sense, instead of talking about imagination in terms of a homogenous faculty, I shall talk of it in terms of capacities, emerging through technical engagement and evolving through individuals and collectives' technical histories.

As a matter of fact, there is a difficulty circumscribing imagination (see, for example, Peter Langland-Hassan 2020, pp. 2–5; Kind and Kung 2016, p. 3; see Chapter 7). Imagination is subdivided into various categories, into a wide variety of imaginative practices. It is also closely related to nonimaginative activities such as perception, memory, desire, belief, hoping, expecting, and so on. But we acknowledge that these boundaries are thin and even require conceptual distinctions which plunge us into scholastic debates without even knowing what we are talking about when we speak of imagination:

> In the case of imagination, there doesn't even seem to be consensus about what the phenomenon under discussion is," much less agreement concerning its deepest nature. In trying to characterize "the phenomenon" of imagination, comparisons are made between imagination and states like perception and belief; but it's emphasized that imagination remains quite distinct from those states. Attempts to specify the precise ways in which it is distinct—and to thereby distinguish what it is we aim to study—threaten to leave us knee-deep in theory, before we've clearly identified what the theory is supposed to be theory *of*. (Langland-Hassan 2020, p. 3, quoting Kind and Kung 2016, p. 3)

In this book, I intend to explain this lack of consensus in a way that shows why it is so important to study imagination as technically constituted. To put it simply, imagination is not a faculty substantially different from other cognitive and conative activities such as perceiving, judging, desiring, hoping, and so on. Making such distinctions between substantial faculties might reflect a referentialist bias, a misleading belief that, when we speak of imagination, we do speak of something in particular, as if the psychological concept, a theoretical construct, by which we categorize a given phenomenon as imaginative or creative, referred to something in the world, something substantial, namely a given imaginative faculty that could be located somewhere, why not in a subjective entity to be studied in a scientific language, for instance that of neuroscience.

However, one could argue that the difference is not substantial or in kind, between imagination and perception, action, desire, belief, and so on, as well as between different forms of imagination (propositional, sensory, experiential; see Chapter 1)—but that the difference is functional. From there, and furthermore, one could argue that this functional difference between imagination and other cognitive activities, as well as between different forms of imaginative activities, arises not only from functional differences in the brain, as it is usually presented (to mention but a very few, Cabeza and Nyberg 2000; Damasio 2001; Duncan and Owen 2000; Dietrich 2004), but from a diversity of technically and socially constituted practices in which these various imaginative and cognitive practices take place, with their corresponding brain processes and dynamical architectures. These functional differences would be a matter of a diversity of technically and socially constituted practices, in which multifaceted imaginative and, more broadly, cognitive acts are enabled, constituted, perfected, and specified, and, finally, acquire a different phenomenology. Again, in the anthropological perspective I advocate, the question is not "What brain is needed for such capacity to exist?" Instead, it is to know "what technical milieu is needed for such brain, with its intrinsic functions, to appear and exercise?"

And there again, such a view on the practical and socio-material dependence of imagination is not new in the history of ideas. Dewey too was eager to make this point: tools enable thought (Dewey 1916). Against any idealistic temptation to talk about thought in the absolute, the pragmatist philosopher urges us to acknowledge the multiplicity of "thought-situations" being opened by specific technical devices and enabling specific sets of cognitive and scientific activities:

> Generalization of the nature of the reflective process certainly involves elimination of much of the specific material and contents of the thought-situations of daily life and of critical science. (Dewey 1916, p. 83)

All the same, talking of imagination in the absolute, without anchoring the reflection in a detailed study of technically and socially constituted imaginative situations, leaves us with the same two pitfalls I criticize in Chapter 2: internalism and representationalism—coined in one word, "neuro-reductionism."

Chapter 5, then, is the place for me to insist, with French philosopher of technics Bernard Stiegler and cognitive archeologist Lambros Malafouris, both explicitly in line with André Leroi Gourhan's anthropological perspective, on the idea that technics enables cognitive and imaginative capacities and acts. As says Malafouris, things shape the mind (2013; see Chapters 5 and 7).

But this leads to a topical difficulty, which I tackle in Chapter 6. To sketch out things, I say that imagination is embodied and ecological, that it has to do with our concrete, embodied engagement with a socially and technically constituted world. And eliminating imagination as a faculty, to replace it with an evolutive set of historically and technically situated capacities, is a theoretical condition for who intends to think what internalism and externalism avoid thinking, namely the constitutive relation between imagination and technics. To put it plainly, I argue that a way for cognitive science to overcome the limitations of internalism and representationalism about imagination lies in reemphasizing, beyond classical cognitive science, the contribution of historical and cultural factors, especially the role of a technical background context in cognitive individuation.

In Part Three, I then evaluate if, and if yes, how we can apply the theoretical choices I unearthed in Part Two, and use the conceptual tools I presented, to think in the language of contemporary cognitive science the technical constitutivity of imagination. I also wonder where such a theory of imagination as technically constituted leads us as concerns the internal or external, and representational or nonrepresentational, nature of images and imagination.

But, precisely, to say that imagination has to do with embodied and worldly engagement faces one classical objection, which is to know how can abstract cases of imagination be dealt with. Ontologically, the ability we have to produce and manipulate images privately ("mentally" in the traditional and internalist sense) might be acquired through practice, by means of technical mediations (using tools, interfaces, instruments, and so on). But, at first glance, the very moment I "possess and manipulate" them, in the intimacy of my closed eyes, those images are neither technical nor external in themselves. One would easily consider with Kosslyn (1994, 2006), for example, that "visual images" consist of both neuronal activation processes located in the visual cortex and private phenomenal experiences. Thus, the question is simple: Why deny that neuronal processes "correspond" to these private and meaningful phenomenal experiences we usually call "mental images"? Again, where are mental images as we privately experience them, if not in a brain necessarily representational? And does this entail that there is a realm of representational images that cannot be brought within the confines of the nonrepresentationalist perspective I offer?

Furthermore, until now, have I properly talked about "imagination"? I talked about material images production, emphasizing the idea that this technical production of material images is an ontic part of the imaginative life; it is the condition of image genesis, as well as of a kind of cognitive enablement that allows us to mentally visualize given musical, pictorial, cinematographic, choreographic, and so on images. But, if I imagine a unicorn flying above my head, or if I tell myself a story, or if I mentally compose, say, a little gentle three-part fugue, without making any movement or noise, am I not disengaged with reality? How could I imagine flying or fighting Yoda with the force if imagination were not essentially a way to overcome the limitations of action and of any possible bodily and technical engagement (Berys Gaut 2003, 2016; Dustin Stokes 2014, 2016)? Who never dreamed away, when "mental images" impose themselves, replacing our conscious experience of reality with fictions, with mental visualizations or auditions of things that are not present in the field of perception?

In other words, what about imagination as we usually define it? How to think, given the anthropological and enactive perspective I develop, that imagination is, indeed, a function of irrealization, a way to evade the rules of concrete action, to disengage with our concrete bodily experience of things, ultimately to escape from reality? What about fictional imagination? And what about this silent musical imagination in the intimacy of which I can sing a given musical piece, and slightly enrich or change it, fulfilling the composer's secrete promises (Gadamer, *Truth and Method*)?

In Part Three, I answer this objection in embodied, ecological, and nonrepresentationalist terms. I combine enactive and ecological approaches with the philosophical perspectives I developed in Part Two. Instead of isolating abstract imagination from human socio-material practices and treating it as if it were the paradigmatic illustration of what imagination is in general, I proceed in the opposite direction. I explain fictional imagination as a particular case of imagination understood as a function of realization and worldly engagement.

In Chapter 6, drawing from my experience as a former professional cellist and from 5E approaches to cognition, I denounce the situations I just mentioned as artificial situations, constructed by philosophers, for philosophers. These situations are better understood as moments in a temporally extended process (musical learning, institutional projects, career planning, and so on) that is intrinsically technical and social. No one imagines "with eyes closed," out of any inscription in a practical, temporally extended, and socially and technically constituted context. Imagination, in this sense, cannot be disassociated from this context and from practical engagement, and then, substantialized or absolutized, to constitute, in the end, such purely artificial situations as if they consisted in the whole imaginative and creative phenomenon.

In Chapter 6, I give particular attention to what researchers of the cognitive science call "experiential imagination." To put it quickly here, experiential imagination refers to the capacity we have to mentally recreate experiential perspectives, with their not only sensory but also motor, proprioceptive and emotive aspects. Experiential imagination is said to play an important role in musical and instrumental learning (Gaut 2007) and imagining (Dokic and Arcangeli 2015). This imaginative capacity enables individuals to imagine performing musical works with a given instrument (Peacocke 1985). A cello apprentice, for example, learns from experiential imagination, by "putting, to use consecrated terms, herself in the shoes" of her teacher (Gaut 2007). Beyond mere sensory images (visual, auditory, and so on), she experiences, in imagination, the affective, proprioceptive, and motor dimension of her playing the cello. She enters in the subjectivity (Dokic and Arcangeli 2015) of a lived, still fictive action or situation.

I focus on experiential imagination and put aside sensorial and propositional imaginations (see Chapter 1) for two reasons at least. First, experiential imagination is of particular interest for the study of musical imagination. Theorists of experiential imagination usually illustrate their intuitions and reflections with musical situations (musical learning, mentally rehearsing a musical performance). Furthermore, experiential imagination allows exhibiting the intrinsic richness of musical imaginative experiences. Experiential imagination suits very well the study of musical imagination on which I focus, drawing from my own personal and professional background as a former professional cellist.

Second reason, the philosophical choice I made to think of imagination as a fundamental mode of engagement with the world, brings it back to its experiential dimension as being as fundamental as it is indissociable from all the other types of imagination which are only subdimensions of it.

This, worth noting, does not amount to substantializing experiential imagination. If we follow what I introduced above, which I shall demonstrate more rigorously in this book, namely that imagination is not a faculty and that the functional and phenomenological differences between different imaginative phenomena are not related to different faculties, but to a cognitive enablement and diversification through technics—then, making distinctions between sensorial, propositional, and experiential imaginations, as contemporary theorists of imagination in the field of cognitive science usually do, is valid at the descriptive level only, but not at the ontological level. Put differently, situating myself in the perspective of enaction, pragmatism, and the French philosophy of technics, I cannot take these distinctions between propositional, sensory, and experiential imagination as a relevant starting point. Rather, such distinctions need to be accounted for, as a result of a technical enablement and diversification of cognitive experience, on the basis of a primitive—biological, social, and imaginative—mode of worldly engagement.

In short, in Chapter 6, I approach what contemporary cognitive scientist use to call "experiential imagination" in nonrepresentational terms. And I define it as a fundamental—biological, social, and technical—mode of engagement with the world. I show that musical imaginative experiences illustrate this fundamental and ecological nature of imagination understood as a biological and technically constituted mode of perceptive, emotive, and agentive experience.

Chapter 7, "Imagination Reconsidered," is for me the place to say positively what I take imagination and images to be. I synthesize the definition of imagination that I progressively, but in a scattered manner, elaborated in Chapters 2 to 6. And I specify how I think the philosophical reflections I offer on imagination, can orientate contemporary scientific research, beyond classical, internalist, and representationalist explanations.

Part One

1

Technics

A Blind Spot

1. Technics and Imagination, Starting Definitions

My conception of imagination as technically constituted, I offer, partly takes root in how French philosophers of technics Gilbert Simondon and Bernard Stiegler, in the wake of André Leroi-Gourhan's work, overcame the limitations of a traditional, instrumentalist, and anthropocentric conception of technics. Traditionally, technics generally refers to all the processes and mediations involved in the realization of an action, of an intervention, or in the production of an object. In this first, etymological sense, technics refers to techniques, that is, ways of doing, of making, of performing, of intervening. Hunting and farming techniques, building techniques, artisanal or artistic techniques, therapeutic and surgical techniques, body and relaxation techniques, computation techniques, mnemonics, writing techniques, organization techniques, even normalization techniques whereby modes of production and use are standardized: as many techniques, among others, which humans transmit, learn, and use. These techniques concretely consist of gestures, postures, procedures, operating methods, and utilization schema. They refer to abilities, many of them being proper to specific technical or instrumental devices. Cellists develop fingering techniques that are different from pianists'. These abilities involve a technical mastery, a know-how (how to do) acquired through lifelong learning and practice. The transmission, normalization, and application of these techniques take place in social contexts, in a cultural and historical milieu that is itself materialized in technical infrastructures and artifacts (Steiner 2010, p. 10; see also Ellul 1977, p. 58; Serris 1994). In this sense and by extension, "technics" also includes all the technical mediations and the technically made objects, devices, and infrastructures that are involved in the transmission, learning, and execution of techniques. Technical milieus are sets of interrelated technical objects (e.g., baroque or modern cellos, harpsichords, forte pianos or modern pianos, midi controllers, scores, and sequencers), infrastructures (e.g., music schools, concert halls, operas, studios, and bars), and socially shared practices (techniques proper to specific schools and styles, learning classes, orchestral activities, and so on). These mediations constitute technical milieus or systems, in which and through which human action and thought take place.

Technics refers, then, to a whole organized and organizing system, constituted by techniques and their technical mediations. As Jacques Ellul (1954, 1977) notices it, with the industrial revolution, the apparition of machines and the generalization of automatization, the term "technics" also came to designate the procedures of machines' construction and utilization. By contrast with mere techniques and technical mediations like tools or instruments for example, technologies (also referred to as "high tech") refer to the complex products of modern and contemporary technical activity, as it takes place in the context of industry and engineering. Technology, in this perspective, is the science that studies these techniques, technical, and technological devices. It refers to the general theory of techniques and technologies, to all the knowledge, practices, and vocabulary specific to a particular technical field. Thus, technics, taken in a general sense, refers to the object of technology, namely the whole system of interrelated technical processes, objects, machines, infrastructures, and socially shared practices involved in human activities.

I will not go into the details of contemporary conceptions of technics here (on this, see Serris 1994; Galimberti 1999). Let me just mention the two pitfalls Simondon and Stiegler urge us to avoid when talking about technics. To say in a traditional sense, as I just did, that technics includes all the technical mediations involved in technical actions, interventions, or productions, is running the risk of adopting an instrumentalist and anthropocentric view on technics, which Simondon and Stiegler denounce as obsolete and misleading. I will come back to this in Part Two. To put it simply here, an instrumentalist conception of technics consists in taking technical mediations in general to be mere means for the realization or materialization of pregiven ends (action projects, satisfaction of needs, materialization of pregiven mental designs). In such a perspective, technics is the whole system of technical means, which human agents conceive, make, and use, in order to achieve ends (action, intervention, performance, production, mastering natural elements, and so on) which they, human agents, design independently from those technical mediations. As I will emphasize in Chapter 4, in this instrumentalist perspective the relation between imagination and technics is a mere relation of *realization*. By means of technical mediations, humans *realize* what they imagine independently from their technical environment.

According to Simondon (see Chapter 3), reducing technics to mere means, which human agents design or posit independently from the technical environment, prevents us from understanding what I intend to explain, namely the role of technics in the very definition or constitution of ends. Technics does not just help humans to realize or materialize pregiven ends or designs (relation of realization). Rather, technics constitutes human's ends and designs. Technics opens up agents' possibilities for action. Technical mediations enable agents to engage in specific, technically constituted experiences—ultimately, imaginative experiences. They transform agents' relation to their environment and to themselves.

By extension, an instrumentalist conception of technics is anthropological, in the sense that it takes technical objects to exist only by reference to human beings. Technical objects, indeed, in the instrumentalist perspective, are said to be conceived, made, and used by human agents, and to serve those agents' intentions. The technical object is an object *for* a human subject. And as such, the technical object receives its meaning and its function (the end it realizes) *from* the human subject. Breaking with this anthropocentric view, Simondon (1958b) and Stiegler (1994, 1996, 2001) suggest we understand technical objects and environments, not as mere objects conceived, produced, and used by subjects, according to those subjects' already-given intentions and rationality. Rather, technical objects and more generally technical milieus, need to be understood as intrinsic and active parts of objectivation processes. Technical objects constitute, or enable, the subject's objectivations, that is, its subjective relations to objects in the world. To put it differently, in Simondon and Stiegler's perspectives—I will explain this more in detail in Part Two—subjects relate to a world of meaningful objects, those objects being technical or not, to the only extent that the constitution of these objects as objects for the subject is technically mediated. Subjects are not de-embedded entities endowed with an already-given power of constitution, filled with already-given representations (representations of how to use an object, of the function of an object). Neither are human subjects reducible to mere biological bodies taken separately from the technical environment (see Chapters 3 and 5). Instead, in Simondon as well as in Stiegler's views, there are subjects to the only extent that there is technics, that there is a dynamic and constitutive relation between agents and their socio-material, especially technical, environment. Subjectivation (to become a subject) and objectivation (to constitute an object) take place in the core of the individual-technics relation. Again, I will develop this way of thinking in Part Two. And by doing so, I shall specify and illustrate my definition of technics progressively.

Then, to start with, I define technics in terms of a technical environment or milieu, including technical activities and mediations, without reducing these mediations to mere means for already-constituted human subjects and ends. This being said, let's see now what contemporary conceptions of imagination tell us about the relation between imagination and technics. As a matter of fact, it boils down to little. Contemporary conceptions of imagination, in the context of cognitive science, give very little attention to technics. One commonly accepted taxonomy (cf. Kind 2016) distinguishes three main types of imagination, namely sensory, experiential, and propositional imaginations. The first major contrast is between propositional imagination and sensory imagination. Propositional imagination does not necessarily imply the possession of sensory images, but simply of sentences, of propositional or syntactic contents. It consists in having attitudes to propositions. We also speak of attitudinal imagination (see Peter Langland-Hassan 2015), that is, attitudes relative to propositions that we can also believe or desire, for example. I can believe or desire that it is snowing two mental states or attitudes, one cognitive and the other conative, which, adequately or not, relate to the world via a proposition that we deem to be "represented" in the mind, in a belief or desire box. In the first case, my belief is correct or not, and in the second case my desire is satisfied or not. The question, then, is to know what an imaginative attitude is and what distinguishes it from other kinds of mental attitudes, like belief, desire, memory, and perception, for example. In any case, in the traditional terms of representationalism, propositional imagination refers to the ability we have to form language-like mental representations of things we do not perceive. It is the capacity we exploit when we imagine, for example, *that* "there is a monster under the bed," a capacity centrally involved in thought experiments, modal judgment, and counterfactual reasoning (see Nichols 2003, 2006, 2009).

As sensory and experiential, imagination basically refers to what can be experienced (see Wollheim 1984; Williams 1973; Casey 1976; O'Shaughnessy 1980; Vendler 1984; Peacocke 1985; Walton 1990; Mulligan 1999; Kind 2001; Currie and Ravenscroft 2002; Martin 2002; Noordhof 2002; Chalmers 2002; Carruthers 2002; McGinn 2004; Goldman 2006a, 2006b; Byrne 2010). Experience, however, is said in several senses. Experience refers in the first place to perception. Imagining the monster under the bed is not imagining *that* "there is a monster." The content of such an imagining is not the proposition "there is a monster," but the vision of the monster, the visual experience I mentally have of this monster. Visual images appear even when one's eyes are closed. In the traditional terms of internalism and representationalism, such private imaginative experiences are thought to be caused by the presence, in the brain or in the mind, of picturelike representations, the so-called "mental images" (Thomas 2014). Kosslyn (1980) asserts the existence of such pictorial representations, or *images*, different from purely discursive or propositional representations. Pictorial representations have spatial properties. As mental images, they represent, Kosslyn and Pomerantz claim (1977), like pictures do. Thus, to imagine is not to perceive (to make the sensitive experience of the thing), nor simply to remember it, since the monster, better not to have crossed its way (Kind 2016). Again, it is not either simply to believe, nor to desire it. Instead, when imagining the monster, we form a mental image of it. In the traditional terms of representationalism, we manipulate a mental representation of it, by means of visual, auditory, olfactory, or tactile imageries.

Sensory imagination, then, refers to the psychological mode of recreated mental states, to the ability we have to mentally recreate conscious sensory or perceptual experiences. Experience, however, is richer than sensation, and exceeds perception. In this sense, sensory imagination is a subdomain of experiential imagination (Peacocke 1985). Some insist on the "heterogeneity of experiential imagination" (Dokic and Arcangeli 2015). Beyond sensoriality, experiential imagination has an emotional and affective dimension (Moran 1994; Walton 1990, 2006; Moyal-Sharrock 2009; Medina 2013). It also has a proprioceptive (Currie and Ravenscroft 2002), as well as an agentive dimension (Goldman 2006a, 2006b; Dokic and Arcangeli 2015).

And finally, as Kant emphasized in his *Critique of judgment* (1790), imagination also can be reproductive and creative. Imagination is reproductive when the person who imagines reproduces images of things without merely recalling them from concrete experience, without believing them either. In the words of Currie and Ravenscroft (2002, p. 8), imagination as reproductive is the capacity we have "to put ourselves in the place of another, or in the place of our own future, past, or counterfactual self: seeing, thinking about, and responding to the world as the other sees, thinks about and responds to it." More than making the absent present, even more than a mere capacity to recall experiences of the body to which modern philosophers like Descartes and Malebranche, for example, tended to reduce imagination, Currie and Ravenscroft's "re-creative" imagination expresses a power of projection into a virtual order of reality that cannot and should not be evaluated with reference to an order of the actual. Imagination, in this sense, exhibits a capacity of virtualization of the self which gives great part of its meaning to the notion of "imaginary" (Cf. Dorothy Walsh 1969 on the foundations of virtual experience in experiential imagination). By contrast, creative imagination is the kind of imagination we observe "when someone puts ideas together in a way that defies all expectation and convention: the kind of imaginative leap that leads to the creation of something of value in art, science, or practical life" (Currie and Ravenscroft 2002).

2. Sensory Imagery without Technics

The classical challenge is to distinguish imagination from perception, in terms of function and nature. On the one hand, perception and imagination intimately relate to each other. Sensory imagery draws its content from sensitive experience. Whether by their phenomenal character or by the neural regions that support them, sensory images intimately relate to sensory experience (Kosslyn, Thompson, and Ganis 2006). In the perspective of simulationist conceptions of imagination for example (Finke 1986; Gordon 1995; O'Craven and Kanwisher 2000; Markman, Klein, and Suhr 2009), sensory images refer to recollections, reinterpretations, or transformations in working memory, of visual perceptions stored in long-term memory. Some even claim that the properties of the content of sensory imaginings correspond, with a high degree of fidelity, to the properties exhibited by sensory experiences themselves (Tye 1991; Sills 2005). When I imagine drinking a cup of tea, my imagining is not a simplification or a falsification of the sensory experience I imagine, but its faithful re-creation. Perception, however, teaches us something about the world as it is, while imagination makes us think what it could be. Perception is resolutely passive, while imagination is at the mercy of the fantasies and desires of who imagines. The passerby passively perceives the external appearance of a building, while the architect solves a design problem in imagination, through visual images he possesses, produces, and manipulates intelligently, mentally (see Sartre 1940, for example).

I will not dwell on this classical debate but just observe, instead, that technics, here, plays no constitutive role. Of course, sensory images participate in technical engagement and reasoning. But they have, by themselves, nothing to do with technical engagement, with the concrete inscription of the individual in a technical environment. As Aristotle emphasized, reasoning in general, and by extension technical reasoning, requires the support of sensory images (cf. Deborah K.W. Modrak 2016, about the concept of *phantasia* and its role in reasoning in Aristotle). Phantasia, appearances, or images, no matter how you call them, are indispensable to cognitive life, both because they direct our attention in action and because they provide our reasoning with pictorial content, making us capable of dealing with objects of the world in their absence. Again, mental images are intentional. Through them we relate to things that are absent. Technics, then, is not in the imagination; it does not constitute the imaginative process. Rather, imagination directs and feeds technical reasonings and oriented actions with sensory images. According to Aristotle, technics exists independently from imagination in the form of a system of exact or inaccurate rules, rules of which Aristotle made the intellectual content of technics defined as a disposition to produce (*Tekhnê*).

Closer to us, clinical psychotherapy and personal development approaches, which have been flourishing since the 1960s and 1970s (see Assagioli 1965; Horowitz 1970b, 1983; Korn and Johnson 1983; Sheikh 2003), take up this way of thinking. For instance, they promote imagery guided techniques helping generate cognitive states in the individual that are conducive to meditation, sleep, openness to others, and so on. Individuals technically manipulate sensory images to initiate, structure, maintain, and nourish their relation to the world and to themselves (cf. the guided imagery products and techniques in Tusek, Church, and Fazio 1997; Willard 1977; Ekstein 2001). On another line, we also speak of imagery-based mnemonic techniques, techniques which consist in memorizing items by means of images and image associations (Bower 1970, 1972; Bugelski 1970; Paivio 1971; Neisser and Kerr 1973).

Even more interesting, and more striking than imagery guided actions, we find, for example, in Van Leeuwen (2016) the notion of imagery imitating actions and the idea that the learning of techniques and thus the acquisition of technical know-how are mimetic:

> Say you want to refer to a woman from last night's party and have forgotten her name. "The woman who walked like this," you say, accompanying "this" with a walking motion that imitates that which you visualize (see Kaplan 1968). This action type doesn't just include imitation of visual imagery; it can include auditory imagery, as when you do an impression of someone's voice, and even tactile imagery, as when you demonstrate a massage technique you learned on a friend. Let's call this type imagery imitating action.

Walking, imitating, and massaging, are techniques, that is to say, technically shaped and oriented actions. And following Van Leeuwen, the action I engage when walking, imitating, or massaging is imagery guided, consisting of the mimetic transposition of a visual, auditory, or tactile mental image in action. Here, it is not the concrete action that constitutes imagination, but the other way round: "imagery can be constitutive of actions." The advantage of this approach to the relation between imagination and technics is that it is theoretically economical. No need, indeed, to speak of imagination outside the traditional sensorialist and internalist framework of its definition. Imagination is and remains sensory and internal. It directs a concrete action that resembles its sensorial content.

3. Experiential Imagination without Technics

Topical questions arise here. One could retort, for example, to Van Leeuwen, that this resemblance, a principle of mimesis between action and imagination, perhaps means that there is more to imagination than mere sensory images. After all, this correspondence between image and action needs to be explained. It is not enough to simply postulate it. One must especially wonder how on earth one can pretend to account for the acquisition of a know-how or technique—walking, imitating, or massaging the bruised back of a PhD student, for example—without action entering the equation. Does one claim that learning to play the cello, for example, consists in watching or mentally visualizing someone playing? And that the imitation is possible without imagination having something to do, minimally, with action, with bodily engagement and, by extension, with the practical, socio-technical contexts in which learned and rehearsed actions occur?

No doubt that imitation serves learning. And Van Leeuwen does not claim to say anything more. But as far as we are concerned, we cannot be satisfied with the notion of imitation without accounting for the isomorphic correspondence it presupposes, a form-to-form correspondence between recollected, transformed, or reinterpreted sensory images on the one hand and concrete, technically shaped actions on the other.

Furthermore, what does technics consist in, in such descriptions? All goes by as if taking imagination by its sensory side only, compelled us to approach technics in terms of appearances only and to avoid reflecting both on its definition and on its constitutive relation to imagination seriously. From the simple point of view of sensory imagery, indeed, there is no technics properly speaking, only the imitation of an appearance or image (*phantasia*) of a given technique (walking, imitating, massaging), imitation of an imitation that leaves in the lurch anyone who wants to study more seriously, in imaginative phenomena, the role of concrete engagement with technical devices and mediations, with socio-technical environments like institutions, scientific, and artistic practical contexts, for example.

The question, then, could be put as follows: by widening the field of imaginative experience beyond mere sensoriality, could we not account for this isomorphic correspondence, for our ability to mimic behaviors on the basis of imaginative processes—and get closer to the idea of a technically constituted imagination?

As I emphasized earlier, beyond sensory imagination, experiential imagination engages us emotively (Moran 1994; Walton 1990, 2006; Moyal-Sharrock 2009; Medina 2013). It also includes proprioceptive imagery (Currie and Ravenscroft 2002) and an agentive dimension (Goldman 2006a, 2006b). As regards the emotional part, imagining myself as Britannicus on the verge of kissing Junia for the last time is not simply seeing, hearing, and touching her with my mind's eye, ear, and lips. Nor is it simply believing it, while knowing it is not true (Cf. Medina 2013). Rather, it is to experience it, in the strong, still imaginative, sense of the term. I want to kiss here. But at the same time, I do not want to, because I know I will never see her again once I have kissed her. When I imagine kissing Junia, I feel carried away by a movement of paradoxical desire, engaged in a set of actions which result in a mixed emotional state of enjoyment and torture. This makes this imaginative experience an experience I feel in first person.

As for the proprioceptive and agentive part, Berys Gaut (2007) shows with force that to learn from the imagination is not simply to imitate an action on the basis of its external appearances only. It also and even essentially is to learn this action *from the inside*. It is to access, beyond the objectivity of an action we perceive when it is performed by someone else, to the subjectivity of this action. When learning the cello, the cello apprentice puts herself in the place of the performer, *as a performer* herself. She does not just target her body with a sounding cello as mere objects of internal visualization. She watches her teacher playing a passage, and figures out, instantly or not, how to reproduce what she sees from the outside. She tries, to take Husserl's phenomenological terms (see the fifth of his *Cartesian Meditations* on analogical apperception) to catch her teacher's intentions and understand a lived and embodied experience, emotional, proprioceptive, and agentive, in her own body and gestures:

> Experiential imagination is an important resource in learning from imagination. This, recall, is a matter of imagining what something is like, and goes beyond

> minimal imagining, involving the sensory or phenomenal presentation of an entertained thought-content. (Gaut 2007, p. 155)

Thus, as Dokic and Arcangeli put it, imagining playing Beethoven's Waldstein sonata on the piano for example, involves at least three types of imaginings: (1) "imagining seeing movements of one's fingers on the keyboard"—which refers to sensory imagination; (2) "imagining having a proprioceptive experience of these movements"; (3) "imagining playing the sonata." Imagining playing an instrument involves recreating the perception of bodily movements, in the form of mental sensory and proprioceptive images or representations (Dokic and Arcangeli 2015, p. 4). It also involves recreating an agentive experience. Motor imagery, Goldman claims, is "the representation or imagination of executing bodily movements," and has its counterpart "events of motor production, events occurring in the motor cortex that direct behavior" (Goldman 2006a, pp. 157, 158, quoted in Dokic and Arcangeli 2015, p. 4).

All the same, a singer does not learn how to sing a given melody drawing only from a visual image of her teacher singing a melody, neither from an auditory image of the sound of her teacher's voice. Rather, in the terms of classical internalism and representationalism about experiential imagination, to such visual and auditory memories she associates the representation of her own bodily and vocal actions. She builds her own bodily and vocal expression in the form of experiential imaginative rehearsals.

By doing so, she answers a question central to musical and instrumental learning and understanding: "how do I produce the same action, or an action approaching my teacher's action?" Arnie Cox interestingly makes this "how" question the primitive and embodied principle of music understanding (Cox 2016). According to him, "mimetic motor imagery" and "mimetic motor action" constitute a bodily grammar by means of which we integrate the embodied and emotional meaning of musical works. We grasp the meaning of a musical element, whether overtly by means of bodily actions, or covertly by means of motor imagery:

> Part of how we comprehend music is by imitating, covertly or overtly, the observed sound-producing actions of performers.
>
> . . .
>
> That implies that my comprehension of such sounds involves simulation of the actions that I infer are likely to have created the sound. Whenever we give our attention to such musical sounds, normally we do not simply hear the sounds, but we also feel something of what it would be like to perform the sound-producing

> actions. That is we mimetically *represent* the sound-producing hand actions, to some degree of fidelity, and such *representations* have an affective dimension, in what it feels like to perform such actions. (Cox 2016, p. 42, emphasis added)

Learning how to play the cello, then, amounts to comprehending from the inside another cellist's musical subjectivity. We *experience in first person*, thanks to experiential imagination, what it feels like to play the cello in such or such way. We enter in the lived subjectivity of a cellist's bowing and fingering. Subjectivity, say Dokic and Arcangeli (2015), refers to interiority, and objectivity, to exteriority. Sensory imagination is purely "objective" in the sense that it takes roots in the perception of the external world. Experiential imagination, in contrast, is subjective for the reason that it is rooted in interiority, that is, in feelings, proprioceptions, and our sense of agency. I will come back to these disconcerting conceptual distinctions and equivalences and to the question of subjectivity in due course (Chapter 6). Let's just consider, here, this idea, widely shared by theorists of experiential imagination, that thanks to emotive, proprioceptive, and motor images, one imagines playing the Waldstein sonata as "her" experience, an experience she feels in her own flesh.

That being said, though, one would legitimately be struck to observe that these descriptions of experiential imaginings remain silent about their eminently technical dimension. It is not without any technique, and out of specific practical, instrumental, and institutional environments indeed, that one undertakes and manages to play or to imagine playing Beethoven's Waldstein sonata. And whether one plays it on a fortepiano or on a modern piano, the technique will change significantly, given the different set of practical conditions and possibilities for actions (affordances, see Chapter 6) each instrument respectively offers. But strikingly enough, apart from a few proponents of enaction and ecological psychology (Hutchins, 1995b, 2010; Malafouris 2013; Gallagher 2017; Van Dijk and Rietveld 2020), to whom I will return in next chapters, this very simple, basic idea that imagining oneself playing the sonata has to do with technics—in some way, at least minimal and to be specified through utter investigation—is nowhere thematized in the literature of contemporary cognitive science!

As I just mentioned, Arnie Cox takes the "how to do it" question to be central to imaginative and interpretive processes. But it does not dawn on him that this question essentially has something to do with technics. Cox's embodied approach to musical cognition naturally gives much importance to action and motor imagery. As a spiritual son of Jean Piaget (1962), Cox maintains that we understand the behavior of others in a mimetic way. The sound image Van

Leeuwen sees us capable of producing when imitating someone's voice is already in itself a motor image. Cox refers to the neuroimaging work of Halpern and al (2004), for example on subvocalization, which shows that the cerebral regions of the supplementary motor area (SMA), that is, the cerebral areas involved in the preparation of movement according to the data of sensation and memory, are activated during the perception and recognition of instrumental or vocal timbres.

But if what "matters," as Cox points out, is the attempt to emulate the sounds, feeling something of what it would be like to be an entity capable of producing such sounds, is it not obvious that this "attempt," this imaginative act, as well as the entity capable of such imaginative acts, might somehow be related, beyond mere embodiment, to technical engagement? Covert or overt, the hand movements, by means of which we understand the meaning of a given prelude, are movements of a pianist or cellist. For sure, Cox duly notes, I don't need to be a cellist in order to grasp the meaning of a cello performance—though musical education does enhance musical understanding. Nonpianists and noncellists possess a kind of gestural grammar, be it primitive, that makes them capable of knowing, for example, what it is like to strike hard or softly in a certain rhythm on a keyboard, or to sway back and forth on a cello. So, by analogy, they get an idea, a feeling, even approximate, of what it could be like to produce such or such sound, such or such arpeggio:

> For example, recall the prelude of Bach's G-major Cello Suite (or some other work for a string instrument that features arpeggiated chords). If you are a cellist you will likely feel something of what it would be like, or is like, to play those arpeggiations, and this motor imagery would be integrated along with the recalled sounds and, likely, something of the notation. By contrast, as a noncellist I have only a very poor idea of what it would be like to play those arpeggiations, and yet I cannot recall or imagine this music without feeling *something* of what it would be like to make those sounds. This mimetic inclination takes the form of subvocalisation combined with an impulse to move my head and torso in concert. (Cox 2016, p. 42)

But whether it be a specific technique or an attempt to find an approximate technique of sound production, the cognitive act, here, still has to do with technics, in some way to be specified. It is an attempt to answer the question "how would I produce," in this case, a sound, an arpeggio, four arpeggiated measures on G tonic. And an "entity capable of producing such sounds," to take Cox's terms—if there is anything like *entities* of this sort—is a technically situated entity. And, if the question "how to produce" is indeed a technical question *par excellence*, then, to situate it, as Cox does, at the heart of interpretive and imaginative processes, induces the necessity of a thorough reflection on the relation of imagination to the technicality of its content, and on the normative and constitutive relation experiential imagination might have to practical engagement with the external, socio-material, ultimately technical environment.

Let me insist. Such imaginative experiences refer to actions, to preparations and anticipations of actions, to a kind of imaginative engagement with action possibilities that are provided by a given technical environment. But the fact that an embodied theorist like Cox, as well as theorists of experiential imagination, even when talking about instrumental imaginative experiences like mentally imagining playing the Waldstein Sonata on the piano, avoid taking into consideration the technical, in this case, instrumental and institutional aspect of such imaginings—this inevitably gives the feeling that technics is a blind spot in contemporary theories of imagination, a kind of theoretical ghost, like a soul in pain waiting to be saved.

4. Creativity without Technics

As I emphasized above, imagination plays an important role in creativity. Creativity underlies a vast array of human practices, from art to science, stage performance, commercial enterprise, and business innovation, for example (Sawyer 2006). It is generally agreed to include two defining characteristics: novelty and value (see Dustin Stokes 2016; Margaret Boden 1990, 1994, 1995; Sternberg and Lubart 1999; Dietrich 2004). Being creative involves coming up with new or original ideas, concepts, and interpretations. Ways of thinking, of producing, of expressing, of executing given actions can be original, that is, unprecedented and unexpected. In such cases, novelty can be relative to the history of the individual or to the history of the collective. A given idea, for example, can be new relative to the psychology of the individual who comes up with it, or, more broadly, relative to the history of ideas. As Stokes puts it, "an x is novel relative only to some comparison class, and this can vary substantially in broadness" (Stokes 2016, p. 258 fn; see also Stokes 2007, 2008, 2011, 2014; Runco 2010).

However, as Kant emphasized (1781, p. 186; quoted in Stokes 2016, p. 247)), "there can be original nonsense." For this reason, in order to be creative, something requires value attribution (Stokes 2016, p. 247). The value, or appropriate character of something novel, does not come from the object that is created, or from the creator and her purportedly creative activity. Rather, creativity lies in how the created object and the processes that led to its production make sense in a given context of sociocultural practices of thinking and making. Then, creativity attribution requires that the x "be of some value to its marker and/or its context of making" (Stokes 2016, p. 247). For example, a technical operative mode is said to be creative when, beyond its intrinsic novelty, it is also useful or adaptive concerning task constraints (Sternberg and Lubart 1999, p. 3).

That said, let me note that Margaret Boden's widely accepted distinction between psychological and historical creativities goes along with the explicit and traditional attempt to explain the former independently from the latter, or to put it more clearly, to avoid thinking seriously the intrinsic relation between the psychology of creativity and the historical context and practices relatively to which creative ideas, actions, and objects get their value. As Boden puts it, "for someone who is trying to understand the psychology of creativity, it's P-creativity (psychological creativity) that's crucial" (Boden 1990/2004, p. 2). In this way, contemporary theories of creativity are mostly internalist, computationalist, and representationalist. Beyond Kant's belief that creative genius was mysterious or even mystical (Simonton 2000), cognitive scientists take pride in originating creativity in ordinary mental and informational processes that are internal to the individual (Boden 1990; Ward, Smith, and Finke 1999; Weisberg 1993; Dietrich 2004). Boden, for example, refers creativity to internal combinatory, exploratory, and transformational processes (1990). Combinatory creativity refers to the new ways we combine preexisting ideas. Exploratory creativity lies in the way we explore the new potentialities of a pregiven conceptual space. Transformational creativity refers to the capacity we have to change the conceptual space, the way we think. These processes consist in the mental-internal manipulation of ideas, of representational contents. And finally, given that "any theory on creativity must be consistent and integrated with contemporary understanding of brain function" (Pfenninger and Shubik 2001, p. 217), the neuroscience of creativity works on revealing the brain mechanisms that underlie creative thinking (Cabeza and Nyberg 2000; Damasio 2001; Duncan and Owen 2000; Dietrich 2004).

I will not detail these approaches to creativity here; instead, I will just observe that, again, the external and socio-material, intrinsically technical world and how we concretely engage with it as bodies in action do not enter into the equation. In the context of internalism and representationalism, which I will define more precisely in the next chapter, material culture (physical objects, behaviors, norms, and rituals embodied in physical objects) merely appears as a way to disseminate new ideas and to recognize their creative value, with respect to a given history understood as a mere record of innovations. The mental, here, is identified with the internal and the representational. P-creativity lies in individuals, in internal and information processes, while history operates as a nonparticipative standard relative to which something new is evaluated afterward, as creative or not (on this, see Ingold 2014; see also Chapter 4). In such a dualistic theoretical situation, the question of knowing whether psychological processes involved in creative activity have some constitutive relation with the historical contexts and practices, relative to which creations get their value, remains unaddressed, unthematized, and unanswered.

As it seems, internalism produces psychologist conceptions of creativity, in the context of which embodiment and concrete engagement with the socio-material world remain unthematized. One could retort to such conceptions of creativity, that making a distinction between two categories of phenomena does not imply that their explanation and ontology differ. We may observe, with Stokes, that creativity is a degree concept, in the sense that "a x novel relative to human history is novel at a higher degree than an x novel relative to some smaller comparison class" (Stokes 2016, p. 247). But if it is a distinction in degree, then it is not necessarily a distinction in nature. So to speak, taking ideas to be novel at different psychological scales, the individual and the collective, does not mean that creativity at a lower degree has nothing to do with the social and historical dimensions that characterize a higher-degree creativity.

Peter Carruthers shares with embodied approaches to imagination a strong rejection of the widely accepted reduction of imagination to thought and sensation. As a representationalist, Carruthers rejects the strong forms of the embodied cognition thesis. Still, he breaks away with the idea that imagining consists in manipulating simple thoughts disconnected from action and practical contexts and that creating consists in assembling such ideas in new ways only. Carruthers presents his act-first account of creativity, as a radical critique of this widely adopted intellectualist reductionism. Against Currie and Ravenscroft, for example, who consider that creative imagination consists in "putting ideas together in a way that defies all expectation and convention," Carruthers asserts that ideas, in the form of visual imagery or speech, emerge subsequently to mental rehearsals of actions:

> According to the act-first account, action schemata can be activated and assembled creatively (in a constrained stochastic manner) without the assemblage being guided by prior creative thoughts or intentions. When this happens the

> resulting action schema is either implemented in action immediately, or it can be mentally rehearsed, issuing in imagery of the represented action. (Carruthers 2011, p. 438)

Thought and sensory images result from a mental and creative play of action representations that are largely concerned with a technical dimension of experience:

> The outcome of a creative process can be an object (whether concrete or abstract), like a new tool or type of tool, a new management practice, a new painting, or a new piece of music. But a creative outcome can also be an action or type of action, like a novel sequence of movements in free dance or a new way of using an existing tool. (Carruthers 2011)

This also goes against another widely shared idea which I shall rigorously refute in the following chapters, according to which, imagination's creativity, freedom, and fantasy lie in its essential disconnection from action (see, for example, Gaut 2003, 2016; Stokes 2014, 2016). Carruthers links creativity and action to each other. He explains the creativity and the play of sensory imagery, as resulting from the semi-random play of structures or "action schemata." As it appears, this amounts to rearticulating imagination to action.

However, Carruthers may not be satisfied with making action an epiphenomenon of the imagination. He sure takes action to be nothing less than the core operation of a certain kind of imaginative and creative process. But still, he defines action in a weak, as a mere mental rehearsal of action schemata, as the resolutely representational constituent of purely internal processes. Carruthers' act-first account of creativity remains unable to account for the role, in imaginative and creative processes, of concrete, socio-materially constituted action. I will come back to this in due course. But this tendency to speak of action in the sense of mental structures of action, in a purely internalist and representational sense, remains an intellectualist sin that Carruthers commits, a sin shared by a number of authors, like Cox, who nevertheless base all their theoretical hopes on the notion of action.

2

Avoiding Representationalism and Internalism

1. Representationalism and Internalism: Genealogy of a Philosophical Sophism

1.1. Substantializing What Cannot Be Substantialized

In order to mark significant progress, a scientific inquiry always needs to question the postulates and conclusions of a preceding one (Dewey 1938). In some measure, scientific progress proceeds critically in avoiding the epistemological traps that skewed the interpretive frameworks adopted in previous scientific inquiries. The stake is to set up new conceptual tools, new theoretical ways to formulate and address given scientific problems, as well as to interpret given experimental results.[1] This is why pragmatists are of such interest for 5E Approaches. They provide powerful philosophical tools to put into question the core epistemological and methodological principles of cognitive science (Gallagher 2016; Steiner 2008a, 2008b, 2010, 2011, 2014, 2016, 2017, 2019a, 2019b). The question is not to know whether cognitive sciences are pseudo-sciences, grounded on outdated, misunderstood, and misused concepts. Rather, it is acknowledging that cognitive science works like any other science. It needs, in order to mark substantial progress, to continuously question the concepts and metaphors that constitute its interpretive frameworks.

And there is this idea, still widely shared among today's cognitive scientists, according to which, when talking about cognitive activities, it is necessary to speak about mental representations. According to a widespread position in philosophy of mind and more widely in the field of cognitive sciences (neurosciences, cognitive psychology, linguistics, and artificial intelligence), the presence in the mind of mental representations explains, for example, how an object—be it the object of a belief, a perception, an imagination, a reasoning—can be present in the mind, categorized and thought by it. Whether the object really exists in the world (the Eiffel Tower), or whether I only imagine it (a unicorn), the mental representation I have of it allows me to think about it, to refer to it. The established formula is that a system X maintains a cognitive relation with an object Y insofar as something in X takes the place of (stands for) Y (Haugeland 1991; Andy Clark 1997).

This view is called "representationalism." Mental representations, very basically, are constitutive entities of cognitive systems which allow these cognitive systems to maintain a relation with an object, in principle independently of the practical, sensory-motor engagement with the world. Symbols, rules, images are mental representations through the mediation of which minds relate to the world. To quote Jean Marie Gallina, who gives a consensual definition, a mental representation is "the smallest cognitive entity endowed with a meaning referring to an object of the world" (quoted in Steiner 2020).

A classical way to present the history of representationalism is to speak of representationalism as a "cognitive revolution," which occurred in the 1950s through people like Edward Tolman (1948), Noam Chomsky (1959), and Donald Broadbent (1954). This revolution would have had the sense of a double alternative, simultaneously against introspectionism and against behaviorism. Human and animal behaviors could no longer be explained, like in behaviorism, in terms of mechanical and direct relations between sensation and action. It was necessary to assert the existence of a representational mediation between sensation and action (Steiner 2020).

The history of representationalism was marked by debates, giving rise to different and often-incompatible forms of representationalism. A major debate was the one concerning the format of mental representations. The so-called imagery debate opposed a pictorialist representationalism (Kosslyn 1980, 1983, 1994; see also, e.g., Hannay 1971; von Eckardt 1988; Tye 1991; Cohen et al. 1996) to alternative, nonpictorialist accounts of mental representations (see Pylyshyn 1973, 1978, 1981, 2002, 2003a, 2003b; Dennett 1969; Sarbin and Juhasz 1970; Neisser 1976; Hinton 1979; Slezak 1992; Slezak, Caelli and Clark 1995; Thomas 1999, 2010). The question was whether representations are propositional or iconic: Are they enunciable propositions or mental images?

Another major debate concerned the structure of mental representations, this time opposing Pylyshyn's computational-symbolic cognitivism to the connectionism of (among others) Churchland (1986) and Smolensky (1986). The question was then to know if we should consider that cognitive representations are made of propositional constituents that can, as such, be arranged, or if they exist in the form of neural networks which do not possess the property of grammatical and syntactic compositionality that cognitivism recognizes in mental representations.

Many differences, thus, are within representationalism. But whatever version of representationalism they adopt, representationalists agree on at least two things. First, cognition always involves the production, use, or retrieval of mental (intracranial) representations, be they symbols, images, or neural patterns of activation. Second, the concept of mental representations is a foundation of cognitive science, without which there would simply be no cognitive science at all. Put simply, the only viable models of cognitive systems should invoke mental representations of any sort. Here is the profession of faith of Jon O'Brien and Jonathan Opie:

> Without representations, the cognitive science would be completely devoid of tools to explain natural intelligence. We would go even further: without representations there is no possible cognitive science distinct from a biological or physical behavioral science. (Jon O'Brien and Jonathan Opie 2008, p. 54, quoted in Steiner 2020)

As regards imagination, representationalism seems inevitable, all the more so that its formulation perfectly corresponds to imagination as defined in folk psychology. It is a commonsense platitude. Imagination is the faculty we have to possess, manipulate, and produce mental images of things that are absent or nonexistent. Instead of perceiving real things, we mentally visualize fictional and fanciful ones. What happens when, in the intimacy of our closed eyes, we imagine a unicorn, enjoy fiction, make up a story, consider counterfactual truths, and so on? According to representationalism, we form particular kinds of mental representations. Mental representations are variously related to—although also different from—other kinds of mental representations, such as beliefs, percepts, memories, desires, hopes, expectations, and so on.

Here is not the place to detail these definitions, conceptions, and debates, in the meanders of such a plethoric literature. I would rather ask a simple question: Is the notion of mental representation—and that of mental image qua mental representation—necessary to study and understand imagination?

This question is a philosophical question, not a scientific one. It calls for conceptual clarification, not primarily for experimental investigation. In this sense, since the beginning of the 1990s, 5E cognition theories have done an important work of reappropriation of past philosophical traditions to criticize those two pillars of representationalism I just mentioned. The last Wittgenstein, Sartre, the pragmatists of the twentieth century, among others, have built up powerful arguments to suggest, for instance, that mental images are not to be found in some place: either in a mind or in the brain, this brain which a current tendency to the most mysterious metaphysical shortcuts often leads us to identify, like Descartes' pineal gland, as the place of realization or instantiation of the mind—in this case, of mental representations and images.

The anti-representationalism at work in 5E theories, basically consists in (1) denying the existence and the relevance of mental representations in the explanation of cognitive processes, notably imaginative ones. It is a matter of saying that in order to explain cognitive processes that are generally explained in terms of mental representations, one does not in fact need to postulate the existence of mental representations; (2) formulating the project of a nonrepresentationalist refoundation of the cognitive sciences (cf., for example, Chemero 2009; Hutto and Myin 2017; Bennett and Hacker 2006).

Now, to start with antirepresentationalism, I would like to present a methodological argument, according to which appealing to mental representations is not necessary to explain cognition. Antirepresentationalism does not consist in denying the existence of a cognitive and subjective life, in denying that subjects represent things, either in the form of mental images, or in the form of inner speech, for example. The antirepresentationalist does not deny the existence of private representational acts, and of representational capacities, by which we imagine, anticipate, relate to the abstract, to the future, to the past, to the possible, to the impossible. It would not occur to any antirepresentationalist to deny that we represent things to ourselves. Denying the existence of such representational acts and capacities would amount to denying cognition itself, to denying the existence of imagination, reasoning, and so on. This would constitute a return to a caricatured form of behaviorism, beyond which the philosophical project of 5E theories is obviously defined. The question is rather, for the anti-representationalist, to know if in order to explain these representational acts and capacities, it is necessary to appeal to mental representations, that is to say to entities situated in the mind or in the brain, entities which would have an intrinsically representational power:

> The fundamental question that inaugurates the contemporary debate between representationalism and anti-representationalism, is the following: should we, or at least can we also speak of representations to describe the subpersonal and unconscious mechanisms and operations that causally make possible the representational acts in which people figure or represent situations, objects, persons, and more generally relate to the world? (Steiner 2020)

In other words, in order to think about representational acts and capacities, must those acts and capacities be substantialized in the form of prior representational entities, located somewhere, in the brain, for example? At first sight, don't we see the trickery, that of a cyclical reasoning (Ramsey 2017; Steiner 2020), whereby one makes the phenomenon to be explained the principle of its own explanation? A cyclical reasoning by which, therefore, one situates in the *explanandum* (what one wants to explain) and the *explanans* (that which explains), the same entity: representations.

1.2. Mental Images and the Mereological Fallacy

The notion of "mental representation" takes root, along with all the dualisms of which our thought is captive, in a double-edged philosophical sophism, in a methodological bias that consists in overrating the abstract results of our analyses. The so-called mereological fallacy (Dewey 1925/1929, 1934) consists in an inversion whereby the abstract results of our analysis (the concepts of mental representation, of mental image, of propositional content, and so on) acquire the ontological status of a condition, and the epistemological status of an explanative starting point. This amounts to projecting in an earlier phase of experience, properties pertaining to a later and more complex phase. It is enshrining those properties as principles or elementary constituents of the whole process, disregarding the temporal and transactional development of experience.

In the case of mental images, this philosophical sophism works as follows:

1) First phase, that of discrimination. It consists in selecting and overrating a specific phase or aspect of experience, where experience refers to the whole set of agent-world transactions. In our case, we overvalue an aspect of the imaginative process. Mentally seeing or hearing something absent from the field of perception, picturing oneself riding a unicorn, enjoying a fiction of which we are the central character, and so on—different imaginative acts, which we take to be primitive mental states, constitutive of specific imaginative activities like sensory, experiential, and propositional imagination.
2) Second phase, that of abstraction. Abstraction consists in isolating the selected phase and separating it from the other constitutive phases of the process. The phenomenal experience we usually call "mental image" is taken to be a sui generis and irreducible mental state (Kind

2016), and becomes ontologically separated from the other phases of the whole imaginative process. Mental images acquire a specific mode of existence, intrinsically different from the rest of the process. Their mode of existence essentially is nonbehavioral (nonobservable, nonexpressive), disembodied, de-embedded, nonsocial, and nontechnical. It is "imaginative" in the fallacious sense that it is nontransactional, that it has nothing to do with concrete bodily and material engagement.

3) Third phase, that of reification, where we convert this substantialized phase into a new and preexisting reality, into an a priori principle of (imaginative) experience. Defined as sui generis mental states, mental images become the only starting point for the analysis and explanation of imagination, as well as its only constitutive operation.

   Mental images, defined as intrinsically representational and meaningful mental visualizations and auditions, are constructed in this way. They refer to an arbitrarily isolated and substantialized phase of imaginative experience. The mind becomes a faculty to spontaneously produce significations, and imagination becomes a nontemporal and nonhistorical faculty to spontaneously produce intrinsically meaningful mental images. The imaginative mind is said to exist independently from the transactional and practical contexts in which imaginative agents continuously engage and whereby significations appear in the world. Mind and imagination are even posited as originary conditions of experience, action, and signification.

4) Ultimately, such an explanative inversion leads directly to reduction, which is the fourth and synthetic phase of the mereological sophism. Reduction basically consists in excluding from the explanation of the process its other constitutive phases. The conceptual construct of mental images, taken as a primitive element, is then posited as being preexistent to the genesis of images—imagination properly speaking—in the form of an abstraction. Mental representations or images exist somewhere, whether in the mind (whatever it is) or in the brain, separated, as such, from agents' bodily engagements with the material, technical, and social world. As a matter of fact, cognitivism traditionally lies on the deliberate decision to de-emphasize the contribution of historical and cultural factors, and the role of the practical and technically constituted contexts in which actions and thought occur.

At this point, two recapitulative ideas to keep in mind:

a) From a functional point of view, reduction consists in isolating a functional aspect of experience, and take it as an early principle of it. This is what happens when we take those phenomenal and conscious experiences we call "mental images," as well as their corresponding neuronal processes, to be conditional to, and the only operations constitutive of, imaginative activities—in Section 2, I will show that Kosslyn's neuronal and representational reductionism happens to be a good example of such a sophistic approach to imagination.
b) Reduction also leads to substantializing a nonsubstantial aspect of experience. The imaginative mind, for example, a mode of expressive and interactive behavior, is erected as a principle of experience, defined in terms of a faculty independent from behavior and action, regardless of the continuity of interaction modes (see Chapter 7).

Against representationalism and internalism, Dewey thematizes the transactional and practical nature of imagination. In his perspective, imagination emerges through experience. It is a function, like the mind, the self, like personality or subjectivity, that emerges "with complexly organized interactions, organic and social" (Dewey 1925/1929, p. 208). Mind and significations are not prior to transactional processes. The mind has nothing to do with a substantial principle of thought which we can locate somewhere in "individuals" or in their brains. The mind appears in individuals. But it "is not as such individual mind" (p. 219). Rather, it is infused and diffused in between interiority and exteriority. It *exists* and individualizes *in the form* of individual's social and technologically mediated transactions:

> The real existence is the history in its entirety, the history as just what it is. The operations of splitting it up into two parts (mind and world) and then having to unite them again by appeal to causative power are equally arbitrary and gratuitous . . .. To give the traits of either phase a kind of independent existence, and then to use the form selected to account for or explain the rest of the process is a silly reduplication; reduplication, because we have after all only parts of one and the same original history; silly because we fancy we have accounted for the history on the basis of an arbitrary selection of part of itself. (Dewey 1925/1929, pp. 275–6)

To be clear, the mind has nothing to do with an independent entity, emerging at the end of a transactional history which would intrinsically be devoid of subjectivity. Again, the mind is not a pregiven center, from which subjective perspective external things or objects get a sense. Significations do not exist in the mind, in the form of mental images or representations afterward projected or realized in the external world, by means of technical mediations. Transactions are not relations between the mind with its mental images and intrinsically meaningful representations on the one hand, and a passive and static world on the other hand (see Chapter 6 for more details).

Rather, the mind is a defining quality or property of transactions themselves. Dewey says, "childhood is the childhood *of* and *in* a certain serial process of changes which is just what it is, and so is maturity" (p. 275). The same applies to imagination. If, Dewey says, we judge its nature from the creation of works of art, [imagination] "designates a quality that animates and pervades all processes of making and observation." The idea is simple. Imagination exists *as such* in the form of transactional processes, of bodily, social, and material engagement processes. And it emerges through the history of these transactions. It "happens when varied materials of sense quality, emotion and meaning come together in a union that marks a new birth in the world" (Dewey 1934/2008, p. 272).

1.3. Do Mental Images Exist Anyway?

What does such a critical genealogy of mental representations mean, as regards the ontological status and nature of "mental images"? We often use expressions like "having an image in one's head" or "seeing something with our mind's eye" (Kosslyn 1978). If I close my eyes and imagine, say, Don Quixote dancing Charleston with Romeo's Juliet, I picture it "in my mind," without anyone noticing it. This mental image ultimately makes me laugh. If I describe this picture to a friend, she might laugh as well, or think I am wasting my time. Anyway, she should be able to picture the same image, at least, an image close to mine in some ways.

We also use to take images, as "mental," to be private experiences. They are private at least "in the sense that we cannot directly observe other people's mental images" (Richardson 1999, p. 9). Kosslyn takes mental images to be "quintessential private events" (Kosslyn and Ochsner 1994, p. 290). I will come to this assertion later, and show, in the lights of the French Philosophy of imagination and of material anthropology, why it might be misleading. Let me just remark here that, despite their private nature, we can describe mental images, share them with others, and compare our mental images with others'. In this sense, as says Goodman, "discourse about images is hardly less intersubjective than discourse about objects. And our talk of images, so central to cognitive psychology, surely seems not to consist of fairy tales but to be serious, significant, and at its best scientific" (Goodman and Elgin 1988, p. 84). We do experience mental images. We do know how to mentally (in the intimacy of our closed eyes) transform, how to compare, how to communicate, and, ultimately, how to share them. At first sight, "mental images" do exist.

However, what are those "mental images"? No doubt that they refer to experiences, phenomenal and private. But they are not tangible. We cannot see a mental image, we cannot hear what a composer is imagining unless she writes what she hears, sings, plays in her mind—unless she instantiates her imaginings through concrete practical, material, and expressive engagement. As Goodman says, "an auditory image makes no noise; and the pain in my toe I can now imagine does not hurt" (Goodman and Elgin 1988, p. 84). Seeing something like a drawing requires putting this drawing before the eyes. But there is no theater, no screen, and no internal eye in the mind to see those drawings. No matter what metaphors say, there are no pictures, no maps, no movies—in two words, no external representations like "images" properly speaking—in our heads.

Again, no doubt that those visualizations and auditions we usually call "sensory images," for example, exist in the form of phenomenal experiences, ultimately resemble external representations like drawings, digital images, instrumental performances, and so on, and correspond to neuronal processes. But why define such phenomenal experiences in terms of "mental representations" and of "mental images"? If mental images have nothing to do with substantial pseudoentities exhibiting the same characteristics as external representations, and if they are not to be localized in an arbitrarily isolated part of the imaginative process like the brain or an abstract mind—in one word, if the mind extends beyond the mere brain and concretely consists of agent-world transactions—again, what those "mental images" truly are? And where are they, if not in the brain or in the head?

For his part, Goodman simply concludes that "mental images do not exist." They are . . . nowhere (1988 p. 84). Instead of making any substantialist confusion, he emphasizes the practical nature of imaginative and phenomenal experiences. Fine. But . . . don't we have a mental image or moving picture of Don Quixote dancing Charleston with Romeo's Juliet? "Yes, answers Goodman, if having that image is construed in terms of my ability to describe or picture or sort out descriptions or pictures of the image" (1988 p. 87). In other words, there are private imaginative experiences only insofar as that there is a capacity to describe and to produce external representations like drawings, paintings, pictures, movies, maps, models, digital images, as well as textual or oral descriptions, instrumental performances, and so on. I have a "mental image" of a unicorn or a centaur to the extent that I can describe or produce external representations (pictures, drawings, animations, written or oral descriptions, etc.) of unicorns and centaurs. In this sense, a so-called mental image has nothing to do with a given elementary and substantial constituent of imaginative experience. Rather, it refers to specific and diverse kinds of cognitive experiences that exist in the form of specific and diverse, concrete, and practical (descriptive, communicative, productive) engagements with external representations, as well as with the tools of their manipulation and production:

> Having an image amounts to not possessing some immaterial picture in something called a mind but to having and exercising certain skills—a matter of producing, judging, revising certain material pictures and descriptions. To Howard Gardner's statement that "in talking about human cognitive activities, it is necessary to speak about mental representations" perhaps we should add a clause: "but talk of mental representations turns out in the end to be talk of cognitive activities." (Goodman and Elgin 1988 p. 90, quoting Gardner 1985)

As we can see, Goodman's move, a move I will prolong and explain further in next chapters, is not to roughly deny the existence of a private imaginative life. Denying the existence of private imaginative experiences would mean, says he, "to condemn an indispensable part of ordinary and psychological discourse on grounds such as would serve, no less plausibly, for condemning the works of Cervantes and Shakespeare" (1988, p. 89). Rather—a classical pragmatist move—Goodman replaces the language of mental representations with that of "cognitive activities." By doing so, he intends to offer to cognitive psychologists a way to talk about private imaginative experiences, without referring to mental representations, which he takes, following Dewey, to be sophistic constructs.

However, it is not just a matter of replacing a formula, that of "mental representations," with another one, that of "cognitive activities." The stake, indeed, is to substitute an ontology to another. Instead of talking about imagination in the terms of an intellectualist and substantialist ontology, a pragmatist ontology leads to situate images in concrete practical contexts. Images are not sui generis mental states, intrinsically separated from agent-world transactions. Rather, they essentially relate to exteriority, especially to objects and external representations. They refer to concrete agent's activities upon objects and events:

> The cognitive psychologist may be glad to hear that there is a way of making sense of talk about mental images, and take comfort in the thought that the

> availability of an automatic process for purifying talk of images leaves him free to go on exactly as before . . .. But the treatment of image talk I have been suggesting is not a quick and easy excision of some pseudoentities; It does not amount to translation by routine application of a simple formula . . . Rather, what goes on is replacement of statements ostensibly about images by statements about objects and events. (Goodman and Elgin 1988, p. 89)

This conditional reference to "objects and events" is essential to the pragmatist perspective I will develop in the rest of this book. Again, as I just emphasized, those "cognitive activities" refer to practices, that is, ways to engage with external things, in particular, with external representations and the tools of their manipulation and production. No need in such a pragmatist perspective, to deny that when reasoning, imagining, remembering for example, agents manipulate private and conscious representations. But these representations are to be conceived as intermediate and temporary internalizations of concrete, environment-dependent practices. Such private, conscious, and representational acts get their meaning, not from intrinsic properties of natural, neuronal processes (see Chapter 6) but from technically constituted and socially shared conventions, norms of use, norms of correctness.

Let me here, in a few words, summarize the very idea I draw from Dewey and Goodman's pragmatist perspective. There is no question of denying that there are phenomenal experiences like private visualizations, auditions, and experiences. The key argument is to say that these private phenomenal and imaginative experiences have no representational property independently from the concrete and transactional practices in which they participate. "Mental images," understand, private imaginative experiences, take place in the course of a transactional history, independently from which there are no representational phenomena. In this sense—I will develop this idea in the next chapters, especially in Chapters 6 and 7 in ecological terms—imagining does not consist in manipulating mental representations—whether linguistic or iconic—of things, those mental images existing somewhere in a hypothetical mind or in a magical, intrinsically representational brain. Rather, if imagining is, indeed, representing (I represent to myself a unicorn and a centaur; I can see them with my mind's eye), the only representational process, here, lies in transactions, in how we engage in concrete practices, in the technically and socially constituted manipulation and production of external representations (written or oral descriptions, material productions like drawings, paintings, and so on). This is to say that the representational and phenomenal character of mental visualizations and auditions comes from specific practices, those practices referring to concrete engagements with material representations, in embodied, socially, and technically extended practical contexts. We can share, compare, communicate images. And the reason for this might be that, as phenomenal, ultimately private experiences, images *exist in the form* of descriptive, communicative, productive practices, that is, embodied, technical, and social practices.

2. Images in the Brain? An Ontological Question

Let us take stock. I adopt a pragmatist view on imagination, according to which imagination always relates to how we engage in practical contexts, those contexts being intimately constituted through social and technical mediations. Imagination is not a faculty we have to mentally manipulate and spontaneously produce sui generis mental states called "mental images," as if mind and imagination were independent from agent-world transactions. Rather, imagination refers to a mode of subjective experience, where subjectivity refers, in pragmatist terms, to a process of sense-making, a process always potentially intersubjective because it only takes place in the core of agents-world transactions. The core idea I will develop in this book can be formulated this way: an ontogenesis of cognition is inseparable from an ontology of social and technical transactions and, ultimately, from an ontology of technical and social mediations, understand, external representations, and the tools of their manipulation and production. Put simply, understanding how imagination appears and exercises in individuals at the same time is a reflection on how social and technical mediations enable, constitute, and transform imagination through the history of agents-world transactions (see Chapter 5).

It is not—I see the easy objection coming—just asserting that mental images always originate in empirical, perceptive, or experiential situations, leaving unanswered the question of how we manage to imagine nonperceived or nonexperienced things. An ontology of the social and technical, in one word "transactional," nature of (imaginative) experience is, of course, irreducible to a pale caricature of classical empiricism. Dewey suggests it, and I will dig into this sense throughout this whole book, a strong theory of experience (*empeiria*) as transactional and historical, helps accounting for creativity and the ability we have to imagine nonperceived things, without necessarily locating their principle in a mind or a brain distinct from experience, disconnected from action.

This being said, how does the pragmatist language I adopted until now help us resist the widely spread idea that "mental images" understood as private experiences exist in the brain, in the form of neuronal activation patterns? Again, we do have experiences of this nature and they do correspond to neuronal activation processes. Using brain imagery, contemporary neuroscientists take mental images to exist in the form of isomorphic neuronal activation processes (Kosslyn, Thompson, and Ganis 2006). Why not believe them? Why should we not consider that imagining consists in brain and computational-like or associative processes that are independent from the external (beyond the brain-skull boundaries) world, from bodily, social, and technical engagement? Why is externalism about imagination necessary, when (1) the mere brain might be sufficient for explaining the phenomenon (economy principle), and (2) abstract imaginative processes, at first glance, do take place independently from any kind of material, bodily, technical, and social engagement?

I tackle this question in Chapter 6. But to close this section, I would like to introduce the main line of argument, which I intend to elaborate in next chapters through material anthropology, the French philosophy of technics and 5E cognition approaches.

2.1. Reading Poetry with Wittgenstein

Neuroscientists assume that the brain thinks, wants, perceives, categorizes, and so on. Crik's "astonishing hypothesis," as he calls it—as Bennett and Hacker observe, "astonishing" might be overestimating, given that this hypothesis is shared by the most orthodox and widely spread view in today's cognitive science, and was already propounded in Epicurean atomism from the first century BC Christ (Lucretia, *De Rerum Natura*), as well as by philosophers of the seventeenth (Gassendi, Hobbes) and eighteenth century (La Mettrie, Diderot, D'Holbach (see Bennett and Hacker 2003/2014, p. 356 fn)—this "astonishing hypothesis," says Crik, is that "you, your joys and your sorrows, your memories and your ambitions, your sense of personal identity and free will, are in fact no more than the behavior of a vast assembly of nerve cells and their associated molecules." As Lewis Carroll's Alice might have phrased it: "You're nothing but a pack of neurons" (Crick 1995, p. 3). For his part, Edelman holds on Emily Dickinson's verse, according to which "the brain is wider that the sky" and "deeper than the sea," and argues that a key move for understanding consciousness is to define it in terms of integrated activities of multiple brain areas. Brain structures achieve consciousness. Another example—theyare countless (for a more detailed review, see Bennett and Hacker 2003/2014, p. 356, and followings)—in a same verve, Collin Blakemore takes neurons to "have knowledge" and to "have intelligence." They "present, he says, arguments" (Blakemore 1977, p. 91).

Against this reductionist and quite optimistic way of thinking—optimistic in the sense that neuronal reductionism, by attributing to the brain representational powers, is at the same time a kind of representational optimism—Bennett and Hacker emphasize together with Wittgenstein the idea that representational and symbolic properties cannot pertain to subpersonal and natural phenomena. Instead, they derive from social and linguistic practices and institutions, whereby external representations like material images, phrases, models, and so on become meaningful—practices and institutions independently from which no explanation of mind is plausible. In a few words, extending the ordinary use of psychological expressions, generally applied to concrete practical, social, and technical behaviors, to brain alone with its neurons makes no sense.

Wittgenstein argued that "only of human and what resembles (behaves like) a living human being can one say: it has sensations; it sees, is blind; hears, is deaf; is conscious or unconscious" (Wittgenstein 1953, § 281; see also § 282–4, 357–61). As concerns consciousness, there is no way to locate it in the brain, to reduce it to complexly organized neuronal firings.

In the same way, Bennett and Hacker criticize "the neuroscientists' mistake of ascribing, to the constituent *parts* of an animal, attributes that logically apply only to the whole animal," a mistake they call "the mereological fallacy in neuroscience" (Bennett and Hacker 2006, p. 22). Mereology is the logic of parts/whole relations. As I emphasized, a mereological fallacy consists in attributing to parts properties pertaining to the whole. It is an epistemological mistake, a confusion whereby a necessary condition acquires the ontological status of a sufficient condition:

> Our point, then, is a conceptual one. It makes no sense to ascribe psychological predicates (or their negations) to the brain . . .. Psychological predicates are predicates that apply essentially to the whole living animal, not to its parts. It is not the eye (let alone the brain) that sees, but *we* see *with* our eyes (and we do not see *with* our brains, although without a brain functioning normally in respect of the visual system, we would not see). So too, it is not the ear that hears, but the animal whose ear it is. The organs of an animal are parts of the animal, and psychological predicates are ascribable to the whole animal, not to its constituent parts. (Bennett and Hacker 2006, pp. 21–2)

Neuronal reductionism is, at the same time, a dualism (see Bennet and Hacker 2003/2014, p. 72). The brain appears as a new version of Descartes' magical pineal gland, an unconceivable articulation point between two substantialized and heterogeneous domains. Decoupled from material engagement, indeed, the mind (*res cogitans*) is located in the brain, only and arbitrarily isolated part of the material world where, surprisingly, the mind itself is said to reside. All goes as if our still great ignorance of, and sincere admiration for, this complex and sublime brain machinery, prevented us from asking the right questions. As if, somehow, we had the strange conviction that something that exceeds our understanding, must necessarily explain everything.

As a matter of fact, in her famous poem "Wider than the sky" (1862), Emily Dickinson uses a specific figure of speech. She uses a synecdoche. A synecdoche is a figure of speech in which a part is made to represent the whole and vice versa. It is mereological. The brain, says Emily Dickinson, is part of the sky, of the sea, of God. Brain and sky contain each other, brain and sea absorb each other, brain and God interpenetrate like sound and syllables. What does it mean, for the brain to be contained, absorbed, and sung by that which it contains, absorbs, and pronounces? Certainly not a reduction of the whole, sky, sea, and God, to the very little brain part. Instead of reduction, here, it is question of participation, of ontological interpenetration. No dualism, but a sense of ontological overflow between nonsubstantialized realms. The poetical gesture, here, consists in pointing to a kind of double incomprehensibility. This poem, indeed, refers to that which largely exceeds our understanding, the sky, the sea, and God, by referring to the very little part they respectively contain, absorb, and create—a very little part, the brain, that in itself exceeds our understanding.

For sure, this poem expresses kind of a mirror relation between the brain on the one hand and the sky, the sea, and God on the other hand. But it does not necessarily bear a reductionist thesis, according to which the brain being as powerful and infinite as the sky, the ocean, and God simply *is* the mind. Following Evan Thompson's reading of this poem, we may bear the weight of this double incomprehensibility without taking the shortcut of reductionism:

> Reading these scientists, one gets the impression that they see Dickinson as having anticipated and given poetic justification for their view that the mind is just the brain or is what the brain does . . .. But this power of the brain is also terrifying—another quality of the sublime—because, as the poem tells us, the brain also easily contains "you." If "brain" stands for "mind," then the mind is not only wider than the sky; it's also wider than you. You might think that you contain your mind, but it's actually the other way round: your mind contains you—your sense of self—and your mind is too vast for you to comprehend. Think of a dream. From the depths of your mind (or brain) a dream is generated, and

> neither you awake nor you within the dream knows how and why this happens. Instead of there being a separate self that contains and controls the mind, the self is something that the mind (or brain) envisions. (Thompson 2015)[2]

Dickinson's poetical intuition, as Thompson's interprets it, suggests that, as minds, we extend, metaphorically speaking, to the scale of God, of the sky, and of the sea. In more prosaic terms, as minds, we extend to that which exceeds the boundaries of the organism and more particularly of its brain. The brain is absorbed as much as it absorbs. It contains as much as it is contained. The mind, then, is not in the brain. Rather, the mind is *in* the relation of mutual absorbance between the brain and that which is not the brain. Hence, to formulate in the language of cognitive science the very simple idea I oppose to neuro-reductionism when reading Emily Dickinson's poem, the mind's representational and symbolic properties cannot pertain to subpersonal and neurophysiological processes only. Beyond the brain there is the body and there is the world. There are social, institutional, and technical practices that constitute, as ontological dynamical parts, the mind. In simple words, summarizing the whole argument of this chapter and this book, the brain and the mind relate to each other only to the extent that the brain continuously participates in meaningful agent-world, socially and technically constituted transactions.[3]

2.2. A New, Pragmatist Ontology for Brain Imagery?

How does this general idea resist brain imagery? As a matter of fact, the brain is involved. And the private experience of having an image of a centaur occurs simultaneously not to processes in the arm, in the hand, or in the external world. It occurs simultaneously to processes in the brain. Those brain processes are even isomorphic to the forms we imagine. Kosslyn shows, indeed, that

> at least thirty-two distinct areas of cortex are involved in visual perception in the monkey brain (Felleman and Van Essen 1991), and probably more in the human brain. Some of these areas (about half in the monkey brain) are topographically organized; *such areas use space on cortex to represent space in the world* (e.g., see Felleman and Van Essen 1991; Fox et al. 1986; Heeger 1999; Sengpiel and Huebener 1999; Sereno et al. 1995; Tootel, Hadjikani et al. 1998). These areas are not simply physically topographically organized, *they function to depict information*. (Kosslyn, Thompson, and Ganis 2006, p. 135, emphasis added)

Why not endorse, together with Kosslyn and contemporary neuroscientists, the idea that despite the evidence that the brain is part of a whole, it still has the lead role in the explanation of subjectivity and so on? Why not believe given simple brain imagery observations, that "the "mind," indeed, is "what the brain does" (2006, p. 134)? If a form is "represented" in the brain in the form of isomorphic dynamical relations, why say representationalism is false?

Crik says the brain "guesses a complete picture from only partial information—a very useful ability" (Crick 1995, p. 57). As a matter of fact, this phenomenon is even a transhistorical principle in music composition. Bach, as well as contemporary electronic composers, acknowledges this phenomenon as a psycho-acoustic effect fundamental in composition. Elimination of objective sounds saturating the sound space does not mean elimination of the sound itself. This principle made it possible for Bach, to write rich polyphonic music for melodic instruments. Some sounds combinations are interpreted by the listener in ways that enrich its harmonic perception. In neuroscientific terms, all goes as if nonwritten and nonperformed chords were perceived by means of brain activity on the basis of partial information. For example, a double string move like D-F followed by a melodic movement B-flat, G, E, C-sharp gives a sense of an omnipresent, still unwritten and unperformed A. In a similar way, when an electronic music composer hears that a saw bass line saturates the sound space in a way she cannot compensate by means of compensator plugins, she usually deletes the saw bass line, at least part of it, knowing that the listener will somehow perceive the nonexisting, but suggested, bass line, on the ground of partial information. In the terms of a commonplace metaphor which I intend to deconstruct, the brain perceives the sound even when it is not played, as a "logical" part of the living and performed sound. There are, according to this way of thinking, "auditory images" in the auditory cortex, and brain activation in concrete and imagined musical performances function in a similar manner (e.g., Mellet et al. 1998; Kosslyn, Ganis, and Thompson 2001). In a few words and despite what Bennett and Hacker say, the brain hears. Then, if one is ready to say that, why not accept the idea that the brain sees, reasons, decides, and so on?

The question, then, is to know how a pragmatist and externalist language could help resisting the idea that "mental images" understood as phenomenal experiences exist in the brain, in the form of isomorphic neuronal activation patterns.

What I would like to introduce at this point, in line with Wittgenstein and as a general line of argument, is a change of interpretive paradigm. The idea is simple. Neuronal processes, as natural and isomorphic processes, are nonrepresentational constituents of representational processes that are intrinsically ecological and technically mediated.

In this perspective, the isomorphy principle is not an effective argument. As a matter of fact, neuronal reductionism and the isomorphic thesis are not new. Wolfgang Köhler, in his time, said that "the physical book and pen are projected on different parts of the retina, and cause processes in the brain's visual area" (see *The Mentality of Apes*, 1917; *Gestalt Psychology*, 1929). A form in the physical world, he said, is "represented" in the brain, by means of internal dynamical relations, those brain relations being identical or equivalent to the relations that constitute the forms present in the physical world. Those objective forms are "represented" in the brain, understand, reproduced in the form of neural patterns of activation.

In his time, Merleau-Ponty found a nonreductionist way to integrate Köhler's isomorphism to his phenomenology, a way which Simondon, like enaction, appropriated. And I intend to appropriate this solution in my own way (Chapter 6). Let me introduce this solution here. Köhler's psycho-physical isomorphism, Merleau-Ponty says, is a kind of physical reductionism, and makes us incapable of thinking—enaction historically takes root here—the central role of the organism's perceptive, affective, and agentive engagement in cognitive individuation. As Merleau-Ponty emphasizes, it is proper to physical structures to be reorganized from the outside only, without the intervention of an internal, active, and self-organizing/orienting dynamism. In a crucial moment of the *Structure of Behavior*, Merleau-Ponty argues that "if the structures of consciousness are useless in explanation, it is because they have their physical or physiological equivalent; and this 'isomorphism' in a philosophy of form is an identity."

But, says Merleau-Ponty, if we want to understand these brain isomorphic processes as "images" strictly speaking, that is, as *meaningful phenomenal experiences*, we need to say where their signification or meaning comes from. These neuronal and isomorphic activations are indeed not just physical processes, neutral to consciousness. And instead of taking them to function as intrinsically meaningful and autonomous processes, or as having intrinsic depictive powers, it might be possible to explain their representational powers in extended or transactional terms. On this path, Merleau-Ponty suggests that mental images, as private phenomenal and meaningful experiences, must be understood as transitory aspects of a chiasmic relation between the lived and affective body on the one hand and the world on the other. Again, as Dewey said before Merleau-Ponty, "an imaginative experience is what happens when varied materials of sense quality, emotion, and meaning come together in a union that marks a new birth in the world" (Dewey 1934). Simply put, the idea is that what we usually call mental images are transitory aspects of organism-world transactions. They have nothing to do with structures or entities located somewhere in the brain. Rather, they are to be explained in terms of properties of acts or transactions that are intrinsically constituted by the organism's practical engagement with the socio-material world.

In such a perspective, neural isomorphic activations do not exhibit the whole nature of "images" well understood. And brain activations are not meaningful entities by themselves. Imagination does not consist in computing on such isomorphic, supposedly representational, relations. Rather, imagination refers to a phenomenon of image genesis and sense-making, where the organism's engagement with the world, through social and technological practices, plays a fundamental role.

Put another way, those isomorphic relations between states in the brain, mentally visualized forms and forms in the world, could be placed in the context of affective and expressive practices that are socially and technically constituted and independently from which those cerebral activations would be images of nothing. "A topographically arranged sensory area is not an image of anything; there are no images *in* the brain, and the brain does not *have* images" (Bennett and Hacker 2003/2014, p. 183). Rather, neuronal activations and private experiences of mental visualization and audition, for example, participate in ecological and practical processes extending across brain, body, and world. And representational acts need to be explained in the terms of such embodied, practical, and socio-material processes. There are representations, as would say Dewey, in the course of a transactional history that is the only existing form of mind and imagination.

I will explain this in more detail in the next chapters, through the case of musical imagination. Let me just insist again, here, that this is not saying that we have no private imaginative experiences. Neither is it asserting that brain processes have nothing to do with those private phenomenal experiences. Indeed, the very forms exhibited by our private or "mental" visualizations and auditions might correspond to brain isomorphic activation patterns. So far, we might agree with Köhler and Kosslyn's observations.

But I deny that such brain activation processes are "images of anything." In a few summarizing words, here are the ideas which I draw from the pragmatist perspective I developed in this chapter, ideas I shall defend and develop further in next chapters: (1) brain activations represent nothing; (2) images are not entities or structures to be localized in the brain; imagination does consist in producing, manipulating, and combining such representational entities; (3) there are no representational acts and capacities independently from the practical (biological, social, technical) contexts through which meaning appears in the world; (4) creativity cannot be thought in terms of a worldly independent computational process on internal representations. Rather, representing is always an external and transactional process, infused and diffused in between interiority and exteriority, where the manipulation of external, technically produced material images plays a crucial role. As Merleau-Ponty suggests, instead of a substantialist and intellectualist, we need a chiasmic ontology of images qua representations. As 5E approaches suggest, this requires a new research program on cognition, and on imagination, where, again, the anthropological and ethnographic study of social and technical mediations becomes indispensable.

Part Two

3

Articulating Life, Imagination, and Technics

1. The Sense of Life

Enactivists historically developed their theories by elaborating on works by phenomenologists like Edmund Husserl, Maurice Merleau-Ponty or Martin Heidegger (Varela, Thompson and Rosch 1991; Thompson 2007; Gallagher 2012, 2017; Gallagher and Zahavi 2020; Käufer and Chemero 2015; Di Paolo 2018). Interestingly, Gilbert Simondon's philosophy of individuation, in particular his insights about life and about the mind-life continuity[1] in his *Individuation in the Light of the Notions of Form and Information* (1958), as well as in his *Lessons on Perception* (1964–1965), on *Imagination and Invention* (1965–1966), and on *Communication and Information* (1960–1976)—both exhibit and overcome the methodological, epistemological, and ontological limitations of Merleau-Ponty's phenomenological treatment. Merleau-Ponty himself acknowledged the relevance of Simondon's work, in two unpublished working notes[2] for his 1959 Lessons at the College de France, entitled *Nature and Logos: The Human Body.*

Simondon describes life as an organizational process. A biological system continues its own genesis, led by an immanent "organizing dynamism" (Simondon 1958/2005, p. 203). As a "modular" kind of amplification (Simondon 1962/2010), the living is a self-generating and self-distinctive system. As an auto-regulative system, the organism continuously maintains its own "functional unity" (Simondon 1962/2010, p. 169) and individuality. Its constituent substructures vary according to a common process of polarization that defines its unity through space and time. At first sight, we could say that living systems, as Simondon understands them, *look like* autopoietic systems, operational rather than substantial, closed in terms of their organization (operational closure), open in terms of their constituent structures (Maturana and Varela 1972, 1980, 1987; Varela 1979, 1997).

Simondon's conception of life forms part of a wider reflection on *individuation*. Simondon's philosophical project is to conceptualize *individuality* in the light of its *genesis* (individuation), in all domains of reality: physical, biological, and psycho-social (or, as Simondon says, "transindividual"). There are two traditional metaphysical ways to approach the notion of individuality, which Simondon rejects: substantialism and hylomorphism. Substantialism defines the individual being as a constant and nongenerated reality, self-evident and self-explicative. Hylomorphism describes the generation of forms, where the form of a being refers to the set of its defining and constitutive determinations. Simondon's inaugural criticism is that substantialism and hylomorphism are committed to the same philosophical sophism, which consists in presupposing, to individuation, a principle of individuation already exhibiting the characteristic determinations of the individual. Whether we talk about a self-determining being or about the generative meeting of an undetermined matter with a determining form, the principle of individuation—the substantial being (*causa sui*) or the form (*causa formalis*)—carries within itself the very definition of the individual. Such approaches confer an ontological privilege to the already-individuated being over the process of individuation. This takes the form of a circular argument: what we want to explain is presupposed prior to its "explanation": the individual is presupposedly prior to its individuation.

But the individual may be more than what we know by simply observing the result of its individuation. And conceiving of individuation in the light of what we partially know about the individual is running the double risk of (1) reducing our understanding of individuation and (2) ignoring what an authentic understanding of individuation may reveal about the individual, about the true nature of individuality, and about the preindividual conditions of its emergence. This is why Simondon replaces ontology by *ontogenesis*. What is at stake is reversing the ontological privilege attributed by metaphysics to the result over operations, to the individual over individuation—and by doing so, to ensure a complete understanding of reality:

> What is an individual? To this question, I answer that we cannot, strictly speaking, talk about the individual, but about individuation; it is to the activity, to the genesis that we need to go back, instead of trying to apprehend the already individuated being in order to discover the criteria by means of which we know if it is an individual or not. (Simondon 1958/2005, p. 191)

This is the reason why talking about *individuation* refers, in Simondon's language, to *transduction* and, above all, to *preindividuality*. Like Jean Piaget before him, Simondon uses the biological and technological concept of transduction and gives it a new meaning, central to his theory of individuation. In Piaget, transduction refers to preoperational thinking, to a kind of mental operation different from deduction and induction. The child is said to reason transductively when making connections between unrelated concrete instances, using neither deductive nor inductive logical laws. Transduction is proper to cognitive processes that thwart the principle of excluded third. Simondon extends this concept to physical, biological, mental, and social operations of individuation. He takes seriously the way transductive processes escape static logic and applies it to ontogenetic processes. In Simondon, transduction refers to (1) propagation (in-formation as amplification process), (2) relational co-emerging of nonpregiven terms, and (3) the quantic nature of individuating operations. Let us see that in detail.

1. Simondon calls "transductive" an iterative process of individuation (be it physical, biological, psychic, or social), through which a given operation propagates gradually within a domain, "by founding this propagation on a structuration of the domain that is realized from one place to the next" (1962/2010, p. 32). Put in another way, individuation is a transductive process of amplification (Simondon 1962/2010, p. 157) and in-formation, in the sense that ontogenesis is a process through which a domain of reality is shaped (in-formation as "taking form") by another one.
2. This relational and structuring (in-formative) process is nonhylomorphic. An hylomorphic relation takes place between two already-existing domains (form and matter; structures; individual and environment), with their own nature, properties, tendencies, and organizational dynamisms. On the contrary, a transductive process is the simultaneous genesis of inseparable domains of reality. Transduction refers to a dialectical process of individuation, which draws on a preindividual and metastable (neither stable nor unstable) state. The individual (the crystal, the living organism, the image, the technical object, or the collectivity) is a transitory phase of this process
3. Simondon says the individuating system is "quantic" in the sense that through individuation it reaches states that are not deducible from an analysis of its previous states. Life is, then, *not accountable from the start in terms of an adaptive relation* between a pregiven organism and a pregiven environment. As Simondon writes in 1954 in an unpublished letter to his supervisor:

> it is necessary to grasp being before it is analyzed in terms of the individual and the milieu: the totality individual-milieu is not self-sufficient; one cannot explain the individual by the milieu nor the milieu by the individual, and one cannot reduce the one to the other. The individual and the milieu both belong to a phase genetically and logically posterior to a syncretic phase that is constituted by a primary mixture. Here we rediscover an intuition from the Ionian Physiologists, and in particular, Thales. (Simondon 1954[3])

Simondon here already refers to his concept of preindividuality as somehow equivalent to Anaximander's (Thales' disciple) *apeiron* (meaning "that which is unlimited, boundless"), which relates to the original and undetermined mixture from which individuals derive. Before individuality, there is preindividuality. And preindividuality makes each individuation possible. "More than unity and more than identity" (Simondon 1958/2005, p. 26), preindividuality is a *metastable* state of tensions between different orders of reality and of magnitude, a state of unactualized potentialities. Metastability, a concept invented in Thermodynamics and used by Norbert Wiener as well, refers in Simondon to the state of a system, whose stability is easily broken by any intake of information. A metastable state is neither stable nor unstable. Simondon defines the system before individuation as a nonhomogeneous distribution of potential energy, whose reality is nonsubstantial but relational.

Simondon analogically applies these three concepts of transduction, preindividuality, and metastability to the case of organic individuation. The living is an affective-emotive and perceptive-active system, and its affectivity polarizes perception and action. Search for food, avoidance of predators, and sexuality polarize perception, and enact different perceptive worlds. But these worlds do not match, and individuation is, through action, the discovery of the dimension through which they do:

> The problem of the action of the living is precisely the problem of the discovery of compatibility . . .. In order to account for the activity of the living, the notion of stable equilibrium must be replaced by that of metastable equilibrium, and that of "good form" by that of information; the universe in which the living acts is a universe of metastability; the initial incompatibility (*disparation*) between the perceptive worlds becomes the condition of structure and operation in a state of metastable equilibrium: it is the living which, through its activity, maintains this metastable equilibrium, transposes, prolongs and sustains it . . .. The living, entering these perceptive worlds, makes a universe of them, amplifies its own intrinsic singularity. The perceptive worlds and the living organism individuate together in the universe of vital becoming.

> Only this universe of the vital becoming can be taken as a true whole system; but it is not given straight; it is the sense of life, neither its condition nor its origin. (Simondon 1958/2005, p. 213)

This "sense of life" refers to the transductive discovery of the dimension of compatibility in a pluridimensional and incompatible (pre-individual) perceptive world. The "paths of the being" (*Ibid.*) emerge through action, as a systemic co-structuration or co-individuation of the individual and of its associated milieu.

Then, the "milieu" is not a pre-given world to which we adapt depending on pregiven dynamisms. Individuation is adaptive, but adaptation is not self-explanatory. Adaptation is what needs to be explained as the outcome of a transductive individuation. It is not an a priori principle. This is the reason why Simondon rejects Kurt Goldstein's understanding of life in terms of "adaptive faculty" and "preferred behaviors" (Goldstein 1934/1995, p. 266)—which Merleau-Ponty inherits as a way to naturalize phenomenal experience. Against the theory of the reflex structure of the organism's adaptive performances and inspired by Uexküll's "Umwelt" or "Phenomenal world," Goldstein refers to the intrinsic orientation of the organism, to its preferred behaviors through which, despite external perturbations, it achieves its own preferred equilibrium.

Goldstein does understand life as a systematic whole where the individual and the environment define each other. The environment is not pregiven, and its meaning is constituted through the living's activity. However, according to Simondon, by giving the organism and its intrinsic and pregiven adaptive dynamism the explanative role, and making this dynamism the very criterium for discovering the essence of life (Goldstein, 1934/1995, p. 286), Goldstein avoids the question of "vital becoming" and fails to grasp normativity and the essential (because genetical) articulation of meaning *in between* ("at the center" as Simondon says) the individual and the (material, social, technical) environment:

> Goldstein did acknowledge the sense of this systematics of the whole; but, by treating it as organismic unity, he was somehow forced to take it as a principle instead of as sense: hence the Parmenidean aspect of his conception of the being: the whole is given at the origin, so that the *vital becoming* is hard to understand as *an effective dimension of this systematic*. The structure of the organism could be better understood at the level of perceptive worlds, in Goldstein's theory, than at the level of the activity properly said. (Simondon 1958/2005, pp. 213–14, emphasis added)

This leads Simondon to rejecting a classical and hylomorphic understanding of adaptation:

> The notion of adaptation represents in biology the projection of the habitual relational thinking with an obscure area between two clearly contrasting terms, as in the hylomorphic schema; incidentally, the hylomorphic schema appears in the notion of adaptation: the living being finds in the world forms that structure the living; the living, furthermore, gives form to the world in order to appropriate it: adaptation, passive and active, is conceived as a mutual and complex influence in hylomorphic terms. However, adaptation being considered by biology as the fundamental aspect of the living, it is quite normal that psychology and loosely structured disciplines, in lack of principles, believed finding in biology an accurate and deep expression of life by using in other domains the principle of adaptation. But if it is true that the principle of adaptation does not deeply express the vital functions and cannot account for ontogenesis, it is necessary to reform all the intellectual systems grounded on the notion of adaptation. (Simondon 1958/2005, p. 210)

Simondon's redefinition of dialectics comes with a new concept of adaptation. The dialectical scheme is classically used to describe adaptive situations in which the organism, given with its own intrinsic tendencies and preferred equilibria (*thesis*), adapts to external perturbations (*antithesis*) by restoring its own positivity. Adaptation is a mere game of tendencies and countertendencies and refers to the affirmation of the organism (the positivity of its own individuality and functional integrity) over its negated or transformed environment. In a more qualified way, Simondon's *transductive* (instead of synthetic) dialectical scheme thematizes the inventive co-emergence and co-definition of the organism and the environment. No ontological privilege is given to the intrinsic organization and tendencies of a pregiven and self-sufficient organism. The tensions and incompatibilities between the organism and the environment (preindividuality), instead of being negated in a synthetic process, are integrated in a broader systematics by means of technical and social mediations. Adaptation is not the affirmation of the individual over the environment, but the systemic and qualitative change through which the individual and the environment define each other, as part of an indivisible system. *Normativity* refers, then, to the individuation of the relational and operative structures constitutive of reality. Thesis (individuality) and antithesis (preindividuality) are part of each other and merge in an inventive process of vital, social, and technical individuation.

2. The Technological Genesis of Affordances

Such a transductive understanding of adaptation as the co-emergence and co-definition of an organism and its environment is definitely in line with enaction (see Di Paolo 2018, 2021; Di Paolo, Cuffari, and De Jaegher 2018). The question now, is to know how to articulate this Simondonian conception of life and adaptation, to a reflection on the articulation between life, imagination, and the external, socio-material, and, especially, technical environment. And here again, a little detour by Merleau-Ponty might be relevant.

In Merleau-Ponty, the affective and pre-reflexive human body experiences objects through "body schema" and "motor intentionality." Through action, the living human being defines a world, giving sense to his co-emerging environment. This is the core idea of enaction. For Merleau-Ponty, human behavior is not to be taken as separate from its social and technical mediations. In *Structure of Behavior* (1942), he writes that the psychic "[integrates]" the vital. "The normal man, that is, integrated" (Merleau-Ponty 1942, p. 180), emerges through the integration of his vital behavior within the sphere of his human and technical manipulations:

> Human work inaugurates a third dialectic. For, between man and the physico-chemical stimuli, it projects "use-objects" *(Gebrauchobjekts)*—clothing, tables, gardens—and "cultural objects"—books, musical instruments, language—which constitute the proper milieu of man and bring about the emergence of new cycles of behavior. Just as it seemed to us to be impossible to reduce the pair: vital situation-instinctive reaction, to the pair: stimulus-reflex, just so it will doubtless be necessary to recognize the originality of the pair: perceived situation-work. (Merleau-Ponty 1942, p. 162)

Meaning is an aspect of the affective relation between a given organism and its environment. To give meaning to objects in the world consists in perceiving them as actualizing affective, intentional, and environmental potentialities. In the case of human behavior, technical and instrumental devices provide the subject with specific possibilities of action. These devices embody affective values and engage the subject in situations which he perceives as meaningful wholes:

> The workbench, the scissors, and the pieces of leather are presented to the subject as poles of action; they define, through their combined value, a particular situation that remains open, that calls for a certain mode of resolution, a certain labor. The body is but one element in the system of the subject and his world, and the task obtains the necessary movements from him

> through a sort of distant attraction, just as the phenomenal forces at work in my visual field obtain from me, without any calculation, the motor reactions that will establish between those forces the optimum equilibrium. (Merleau-Ponty 1945/2012, pp. 108–9)

The same applies to imagination when technologically or instrumentally mediated. Imaginative processes involve affective and kinematic intentional structures embodied in our manipulative relation with instrumental devices. The organist, for instance, imagines through his instrument, which offers him specific intentional, imaginative, and expressive horizons:

> During the rehearsal—just as during the performance—the stops, the pedals, and the keyboards are only presented to him as powers of such and such emotional or musical value, and their position as those places through which this value appears in the world. (Merleau-Ponty 1945/2012, pp. 146–7)

The instrumentalist never just moves his physical body or refers to it objectively. Rather, he is affectively and pre-objectively conscious of what his instrument requires and "obtains" from him. The instrument polarizes both his gestures and the musical and motor intentionality constitutive of his musical imagination. The "melodic character" of the musician's motions (Merleau-Ponty 1945/2012, p. 107) as well as its direction and coherence, all lie in the way his body is emotionally affected by his instrument.

Like Merleau-Ponty, proponents of enaction emphasize the affective basis of embodiment as key to a nonrepresentational type of intentionality underlying perceptive, imaginative, and manipulative processes (Hutto 2007; Bower and Gallagher 2013; Gallagher 2014, 2017). Hutto suggests the existence of a prelinguistic "kinematic imagination" involved in instrumental manipulation and crafting (Hutto 2007, p. 76). Gallagher's affordance-based conception of imagination also shares Merleau-Ponty's intuition about the intentional, affective, and, in some cases, manipulative relation between the subject and its cultural environment. In Gallagher's definition, imagination can involve the manipulation of toys, props, artifacts, instruments, and so on:

> Imagining involves a variety of different practices—some of them actively embodied, some of them involving the manipulation of bits of the environment, some of them sitting still and picturing something by manipulating concepts or thoughts or images (re-enacted perceptions)—which in any case may still involve affective and kinaesthetic aspects of embodiment. (Gallagher 2017, p. 195)

Affectivity makes the subject an open and situated system, a perceptive, imaginative, and manipulative system connected to its material environment. Walking in Goldstein and Merleau-Ponty's footsteps, enaction, then, straddles the transitive and the intransitive meaning of the French verb *vivre* ("to live"). As Renaud Barbaras emphasizes, "in the French language, the verb *vivre* means both 'to be alive' (*Leben*) and 'to have an experience, to feel something' (*Erleben*): it is neutral with respect to the distinction between the transitive life that we call consciousness, and the intransitive life of organisms that merely keep themselves alive" (Barbaras, 2010). This neutrality reveals the phenomenal and intentional status of life in Merleau-Ponty and points to one of the core ideas of the enactive approach to life and subjectivity. The world we live in is a world we enact, our own world of lived and subjective experience.

I will come back to the notion of subjectivity in Chapter 6. For now, by subjectivity I basically refer to what makes an experience conscious and meaningful. As Myin and O'Regan put it, subjectivity includes the feeling that a given experience is *my* experience (or yours, or his or hers, or ours) and the feeling that something I experience is *for me*, offers me an opportunity to act or think (Myin and O'Regan 2002, p. 30; see also Thompson 2007). In this respect, for the enactive approach as well as for phenomenology, the theory of intentionality, as I just presented it in terms of organisms' affective, world-directed (intentional), and action-oriented relation to the environment, is central to the definition of subjectivity (Thompson 2007, p. 15).

That said, two intertwined questions arise. How is it possible, first, for cultural objects to display affective, intentional, and expressive values? According to Merleau-Ponty, the organ displays affordances, in other words possibilities of action, and expressive values. Going back to the example I mentioned in Chapter 1 when talking of the instrumentalist's experiential imagination, when imagining playing Beethoven's Waldstein sonata on the piano, the pianist engages with expressive values that are mediated and afforded by the piano. She imagines as a situated agent, daily dealing with meaningful and affectively charged keyboards, pedals, and so on.

But where do these values and affordances come from? How is to be understood the articulation between the very structure of the object on the one hand and its values on the other?

Subsequently, a second line of questioning bears upon how the subject can perceive these values. How is it possible that the subject identifies particular uses in the technological or instrumental devices she manipulates? Specifically, how are we to understand, according to Merleau-Ponty's famous formula, that "the world is inseparable from the subject, but from a subject that is nothing but a project of the world" (1945/2012, p. 454)—in this case, a technical and instrumental world? For the postulate, according to which technical objects have a meaning, is not self-explanatory. It remains necessary to account for it, without assuming that the subject already knows from the start how to use her technological surroundings. Otherwise, we would be guilty of the same explanatory lack I denounced in Van Leeuwen's representationalist theory of imagery imitating actions (see Chapter 1). We would not be more capable than internalist theorists of experiential imagination to explain in which sense imagining playing Beethoven's sonata on a piano has to do with technical engagement. How are we to think, as I put it in Chapter 1, of the normative and constitutive relation experiential imagination has to practical engagement with the external, socio-material environment?

It is not enough to say that technics assists living organisms, enhances their capacities, or provides new intentional, affective, and imaginative horizons, and new expressive rules. Moreover, rather than just postulate like Merleau-Ponty and representationalists the dialectical relation between the subject and its technical world, we need to explain it in the light of its genetic and essential relation to life and technics.

This is precisely the core of Simondon's philosophical project, and exactly what the enactive approach has recently made it a priority to achieve. For this reason, it is not surprising that proponents of enaction have founded their reflection about technics either on Simondon's work or on philosophical works related to Simondon's theory of individuation and technique (Stiegler 1994, 1996; Ingold 2013). In a precursor paper, Havelange, Lenay, and Stewart (2003) combine the enactive outlook with ideas of Simondon's and Bernard Stiegler's, one of Simondon's main philosophical heirs. They claim that the apparatus by which structural coupling between organisms and their environment is accomplished includes various kinds of technologies. Technologies mediate the concomitant bringing forth of the organism and its world. On a similar note, Poulsgaard (2019) explicitly draws from Stiegler's phenomenology of technics and Simondon's conception of individuation, as well as from Lambros Malafouris (2013, 2014, 2015), who is a careful reader of Stiegler. Assuming that "an enactive view of mind allows for a better understanding of digital practice by advancing a dynamic, transactional, and affective framework for the analysis of computational design" (p. 1), Poulsgaard regards computational tools and artifacts as essential to cognitive and creative processes. Martinez and Villanueva elaborate on Tim Ingold's concept of "correspondence," which Ingold expressly forged in the light of Simondon's "transduction" (Ingold 2013). They highlight the idea that a type of intentionality, embodied and displayed in the affordances of materials and artifacts, is socially constituted and shared. "Artifactual intentionality" refers to the way technological norms of action, embodied in artifacts, individually and collectively shape creative agents' action-perception patterns.

The idea these developments have in common, central to Simondon's perspective, is that instead of taking the subject's ability to perceive the meaning of technological objects for granted, one must explain it in terms of a prior and historical *genesis of technological affordances in the manipulative and inventive relation between the living being and its technological environment.*

3. Simondon's Conception of Individuation and Technics

Let me now elaborate on this and show how Simondon can nourish the enactive approach to the articulation between life, imagination, and technics. Simondon was a student of Merleau-Ponty. They share a common philosophical tradition and a common thematic horizon. Building on Merleau-Ponty's stance on cultural integration, in a very enactive fashion, Simondon understands psychic individuation as a vital and adaptive phenomenon, ultimately extended into a social and technical environment. As I emphasized, more than a mere transformation of the environment, adaptation refers in Simondon to the co-emergence of the individual and its environment. In order to adapt, the individual does not impose the law of its own intrinsic, pregiven, and static tendencies onto the environment. The enactive principle of co-emergence, according to which mutually constraining domains dialectically emerge and shape one another as part of a dynamic individuating process, resonates with Simondon's conception of individuation. The individual and the milieu mutually shape each other. Adaptation engages organisms' ability to change, and to invent the internal and external conditions of their viability.

Furthermore, and this is where the originality of Simondon lies, he explains this adaptive process in terms of "imagination" and "images." In Simondon, images are pre-reflexive motor structures and spontaneities, patterns of action through which organisms virtually explore their future behavior and grasp the meaning of their perceptive worlds. Like Merleau-Ponty (1957–1958, pp. 248–376), Simondon refers to ethologist Konrad Lorenz's notion of "imprinting" or *Prägung* (Simondon 1965–1966, pp. 33, 93–4, 97), which is a mode of acquisition. Both Merleau-Ponty and Simondon consider it essential for the understanding of the articulation between animality and humanity – or, as Merleau-Ponty says (after French anthropologist Levi-Strauss) in *The Visible and the invisible*, between the "savage mind" and the "cultivated being" (1964, pp. 175–6, 213). As is seen with Lorenz's geese, imprints are acquired patterns of relationship between instinctive behavior and external stimuli. Organisms acquire irreversible instinctive patterns of action-perception that define their worlds in terms of possibilities of action and affective values. A gosling learns the "image of "being a parent"" (Simondon 1965–1966/2008, p. 94), for instance, in terms of a relational scheme with its gaggle that requires specific environmental answer signals – which ethologists may learn in order to study animals from within their natural environment.

As Merleau-Ponty highlights, these instinctive tendencies or action-perception structures "are not actions directed toward a goal, not even toward a distant goal of which the animal is aware. Instinct is a primordial activity 'without object', *objektlos*, which is not primitively the position of an end" (1957–1958, p. 190). Strictly speaking, instinctive tendencies engage living organisms even in the absence of their referential object. Simondon expands this aspect of Lorenz's theory to imagination in general. It becomes a tool to conceptualize the *biological basis of imagination as anticipation*, as well as the empirical evidence that images, understood as motor tendencies and structures of expectation, constantly underlie perception and action:

> Lorenz and Tinbergen showed that hereditary coordinations are not necessarily reactions to real objects or situations; if the motivation is strong, a weak external stimulation is sufficient, and if the motivation is very high, there is no need of any stimulation in order for the instinctive program to take place. Finally, after triggering by a stimulus, the instinctive action continues to run in virtue of a completely endogenous order. This action just lacks the taxic elements that adapt it to the real object when it is present.
>
> These discoveries are important for research on the origin of movement images because they show that the organism possesses a reserve of complex behavior patterns able to be activated endogenously, when motivations are sufficient; thus, there is a true biological basis of the imaginary, prior to the experience of the object. (Simondon 1965–1966/2008, p. 33)

Simondon's images are pre-reflexive and primitively embodied; they refer to the action-perception patterns whereby organisms affectively engage in perception and action. More precisely, Simondon identifies three different categories of images operating in a cyclic relation: the a priori, the a praesenti and the a posteriori. "Embryos of motor and perceptive activity," motor images play freely, independently and prior to the restrictive perception of objects. At this stage, the first stage of what Simondon calls the cycle of images, "pre-perceptive" (a priori) images provide pre-adaptive "patterns of response to (environmental) stimulations" (Simondon 1965–1966/2008, p. 19). At the second stage, that of perceptive and motor experience, those pre-adaptive motor images stabilize according to the environment. They "become a mode of reception of the information coming from the milieu" (Simondon 1965–1966/2008). Preperceptive images individualize in perception and become "intra-perceptive" (images a praesenti). At the third stage (a posteriori), images become memories or "symbol" (Simondon 1965–1966/2008). They allow the insertion of mental activity both in the world and in the past, as well as the reference to past lived experiences and situations that no longer exist. Through the affective, emotive, and cognitive resonance of experience, images organize "according to a systematic mode of association, evocation and communication" (Simondon 1965–1966/2008). As such, images become "trans-perceptive," in the sense that they acquire an evocative power irreducible to the context of their particular emergence. They help stabilize, decontextualize, and communicate a given experience.

Consequently, to take action in the world is always to anticipate both the act of perceiving (sensory-motor spontaneities involved in the act of perceiving) and the perceived object (Simondon 1965–1966/2008, p. 16), by means of sensory and motor images richer and freer than perception and action. Essentially imaginative in themselves, perception and action fit in between interiority and exteriority, and between endogenous motor and pre-perceptive tendencies and spontaneities on the one hand and information coming from the outside world on the other. More than a mere irrealizing function as opposed to the realizing function of perception, and in no way disconnected from action, imagination is, in Simondon's perspective, the insertion point of the individual in the world and of the world in the individual. Through "intra-perceptive" images, namely images present at the core of perception and action, imagination is a power of realization, actualizing affective and environmental potentialities:

> Imagination as anticipation is no longer a function that emancipates itself from reality and develops in the unreal or the fictive: it initiates an effective activity of realization . . .. The modality of the imaginary is that of potential; it becomes that of unreality only if the individual is deprived of his access to the conditions of realization. (Simondon 1965–1966/2008, p. 56)

However, realization comes with its own challenges:

> What situation does invention correspond to? To a problem, that is, to the interruption, because of an obstacle, of a discontinuity playing the role of a dam, of a continuous accomplishing of an operative project. (Simondon 1965–1966/2008, p. 139)

Invention corresponds to the fourth stage of the cycle. The affective and imaginative resonance of perceptive experience results in a progressive "saturation"—to wit, a kind of internal incompatibility. Invention, moreover, consists in the discovery of the synergic conditions of affective and imaginative compatibility:

> When the subject is separated once again from the object, the image, enriched by cognitive input and integrating the affective-emotive resonance of experience, becomes symbol. From the universe of internally organized symbols, reaching a point of saturation, may result invention, which is the implementation of a more powerful dimensional system, able to integrate more complete images according to the mode of synergic compatibility. (Simondon 1965–1966/2008, p. 3)

Action is said to be "intrinsically incompatible" when there is tension between the sensory-motor subsets and "operative modes" that constitute it. Consider, for instance, the conflict that may occur between incompatible hands and fingers' moves necessary for the organist to perform a given melodic line. Conversely, the incompatibility is "extrinsic" when the individual's behavior is ill-adapted to adverse surroundings, for instance when a rockfall completely impedes one's journey. Invention appears, then, as the discovery of the more efficient and adaptive ways for restoring the continuity of action. In the organist's case, finger technique is an adaptive invented process. It individualizes the operative modes and schemes, restores the continuity of action, the fluidity, precision, and efficiency of the series of instrumental and musical gestures. As for the rockfall, invention may consist, for example, in making a detour, in associating many people to clear the road or in crafting a hydraulic winch with which, despite her limited muscular power, the individual may be able to move the rocks out of her way. In all cases, invention refers to a qualitative change in the operative system, restoring the compatibility between the constitutive operative (sensory-motor) subsets of action as well as between action and the environment:

> Invention is the appearance of the extrinsic compatibility between the milieu and the organism and of the intrinsic compatibility between the subsets of

> action. Detour, instrument crafting, collective association are different ways to restore the intrinsic and extrinsic compatibility. (Simondon 1965–1966/2008, pp. 139–40)

From this angle, meaning is the imaginative discovery of new operative modes and of compatibility. It is imaginative in the first sense that imagination is the operating mode through which organisms relate to themselves and to the world. Action-perception is imaginative, just like its cognitive and affective resonance. And it is imaginative in the second sense that it is *inventive*. Indeed, restoring compatibility involves the discovery of new operative structures that are not deducible from previous operative structures:

> Solutions appear as continuity restitutions allowing the progressivity of operative modes, according to a progression previously invisible in the structure of the given reality. (Simondon 1965–1966/2008, p. 139)

Meaning consists both in this qualitative change in the individual-environment system, whereby the adaptive conditions of compatibility are restored, and in its affective and cognitive resonance. It is worth noting, however, that in Simondon technics is not, as is the case for instance in Stiegler, a primary condition of psychic individuation and signification. Technical objects complement the biological and social resources of individuation:

> If we consider the created object as a mediator of the relation between living beings and their milieu, it is less difficult to find the link between invention in animal species and in man; indeed, the use of instruments is quite rare in animals. But there is no requirement to consider the construction and manufacturing of instruments as the principal opportunity of invention; the instrument and the tool are nothing but a relay in the creation of objects, one more mediation between the created object and organisms that create it. (Simondon 1965–1966/2008, p. 188)

Individuals elate to their milieu in a process of imaginative and inventive individuation, and the invention and manipulation of technical devices are part of individuation processes and of the emergence of significations. In the case of music, the cellist or the organist's finger techniques, understood as adaptive individualizations of the operative modes constitutive of their instrumental and musical performances, are the inventive and technical conditions, as well as the ontological instantiations of musical values and meanings.

4. Solving Merleau-Ponty's Equation: Simondon's Practical Schematism

Going back to the question I broached in Section 2, how are we to understand that the subject perceives values and affordances in technical objects? How can we account for the articulation between the structure of the object (in this case, the organ) and its intrinsic values? How is it, to quote Merleau-Ponty, that the organist is able to perceive stops, pedals, and keyboards as "powers of such and such emotional or musical value," and "their position as those places through which this value appears in the world" (Merleau-Ponty 1945/2012, pp. 146–7)?

In Simondon's perspective, this ability results from both a manipulative and an imaginative genetic process, which he calls "practical schematism." Musical values denote the affective and cognitive resonance of invention understood as the imaginative discovery of (1) intrinsic compatibility between the organist's operative modes and (2) extrinsic compatibility between the organist, the instrument, and acoustics. In this sense, the crafting and behavior of the instrument ontologically belong to the musical image and value. The organist perceives the affordances and expressive values of the organ since he invented them in a technically mediated process of invention which lies at the core of musical practice.

Uses and values are not given from the start, out there in the objective world, as if objects possessed and displayed their own meaning independently from organisms' imaginative and inventive activities. Nor are they to be found from the start in those organisms, as if meaning and intentionality existed before individuation, manipulation, and invention, as a priori conditions.

Rather, uses and values are invented in a process of technical appropriation, through concrete practical engagement with technical objects. This is the core idea of Simondon's practical schematism. Creativity, as Simondon defines it in *Invention in Technics* (1976, p. 336), consists, for example, in "replacing usual uses with virtual and unusual functions." We creatively appropriate objects, give them new functions to the extent that, faced with new practical and concrete situations, we imaginatively explore new ways to use them. In other words, there are no pregiven uses or affordances displayed in objects. Similarly, learning how to use given objects consists not only in using them and passively waiting for them to shape our intentional structures, and to constrain our body movements. Rather, we learn how to use them by progressively and transductively (1) discovering the operative modes through which we establish, by their means, the intrinsic and extrinsic compatibility—adaptation as invention and sense-making—and (2) individualizing them.

Instrumentalists progressively and transductively invent their musical imaginaries. They individualize their instrument, collaboratively with the instrument maker or not, and their own behavior along with it. The pianist may not be an instrument maker. However, any advanced pianist vividly pictures himself bending over his piano, adjusting every single hammer so it strikes the keys cleanly. Musicians spend hours tuning their instruments. The oboist or the saxophonist, for instance, understand the way their instruments work and carefully adjust their reeds so as to get a specific kind of reactivity and color of sound. The musician individualizes his instrument to make its operative structure the concrete realization of specific expressive and musical values.

And this process of individualization, both of the instrument and of the instrumentalist, is also the individualization of expressive values. The cellist and the cello (as a physical and technically made object as well as a set of affective, imaginative, and expressive values) co-emerge in a process of imaginative and technical individuation. Moreover, the ontological chiasm between mind and matter, between the flesh of the cellist and that of the cello, is the final step of an imaginative and inventive process that only technological or instrumental devices can mediate. Ultimately, the cellist and the cello are entangled to such a point, affectively, viscerally, and operatively (technically), that stealing or breaking the cello amounts to stealing or breaking part of the cellist's soul, part of her world. In this sense, the instrument is not only constituted through an embodied imaginative process. Instead, the instrument is a constituent of the instrumentalist body. The instrumentalist does not relate to her instrument as something external, something to which she relates with a given and substantial body. Rather, the individualization of both the instrument and the instrumentalist's expression is in itself the enaction and individualization of a new kind, instrumentally constituted embodiment.

In Simondon's outlook, this practical schematism results from the imaginative play of "technical schemes." Distinct from symbol images, technical schemes are body schema individualizations. They refer to an imaginative integration into one's own body schema of a given object's operative modes. In other words, understanding how a technological object works amounts to embodying it, by transposing its operative modes onto our own operations:

> A concrete image of movement always involves to some extent a reference to the subject's body schema. Having a concrete intuition of an object's movement,

> amounts to some extent to putting oneself in its place and situation, as if our body was this object. (Simondon 1965–1966/2008, p. 41)

Schematism in Simondon refers to the imaginative process through which we become the objects of our experience. We grasp their essential nature by means of an imaginative and pre-reflexive kind of analogy. Thus, "a child who plays is not only a driver and a rider, but also a car and a horse; the body schema extends to the inner animation of the everyday objects immediately related to behavior. The motor intuition, in the form of behavior anticipation, realizes an implicit animism" (Simondon 1965–1966/2008, p. 42). Without hastily assimilating both views, we can point out that Simondon's understanding of imagination strikingly resonates here with Gallagher's enactive conception of imagination as pretense:

> In the case of children's pretend play, it's not that the child first imagines X, and then playacts it out: rather, the imagination is accomplished in the playacting. Ryle's example: the child can pretend to be a bear. In this case the child "roars, he pads around the floor, he gnashes his teeth, and he pretends to sleep in what he pretends is a cave" (1949, p. 243) . . . the imagining just is the playacting. It's literally enacting something in bodily movement that may include the use of props. (Gallagher 2017, p. 193)

For his part, Simondon extends this body schema reference to the imaginative process of technical invention and transformation. He takes the invented technical object to be the realization or concretization of an inventive technical thought, which he defines as an embodied-imaginative schematic play restoring the intrinsic compatibility of technological operative modes schematically and bodily integrated. Schematism even allows us to anticipate the future unity of technical objects, and to organize the present accordingly:

> The unity of the future associated milieu, in which the causal relations will take place allowing the functioning of the object, is *enacted*, *played* in the same way a role can be played in the absence of the true character, by the schemas of creative imagination. The dynamism of thought is the same as that of technical objects; the mental schemas react to each other during invention (he uses the word for "invention", but does he mean "imagination"?) as the various dynamisms of the technical object will react to each other in material functioning. (Simondon 1958, p. 71, Simondon's emphasis)

For instance, the oboist adjusts her reeds in order to get a specific behavior from them. She embodies the operative structure of her instrument, transforms it, and finds technical ways to establish the continuity between the structure of her instrument, the acoustics, and the breadth of her musicality. The acoustics of the hall, of the church, and of the classroom are all dissimilar. The instrumentalist takes these differences into account. In such a situation, the concrete genesis or individualization of her instrument is simultaneously the genesis or individualization of expressive powers and values.

Simondon calls the schematic identity between the operative modes of the subject and of the object "isodynamic":

> Man, an interpreter of machines, is also the one who, by means of his schemes, established the rigid forms allowing the machine to function. The machine is a human gesture, displayed, fixed, made stereotypy and power of recommencement.
>
> . . .
>
> Between the man who invents and the working machine there exists a relation of isodynamism. . . . To invent amounts to making one's own thought work just as a machine can work, neither according to causality, too fragmentary, nor according to finality, too unitary, but according to the dynamism of lived functioning, accessed because produced, accompanied in its genesis. (Simondon 1958, p. 138)

"Technical mentality" (1961), that is, the imaginative possession and manipulation of technical schemes, originates in the schematization of technical and nontechnical operative modes and relational dynamisms, grasped in the texture of physical, biological, human, and nonhuman reality. Technical knowledge concerns not only the object itself but also its functional and operative systematics, and its constitutive dynamic and operative relations. More precisely, the "historical" scheme refers to the stable scheme of an individualized object, for instance the scheme of a particular ball check valve. The "lineal" scheme refers to what is common to the technical lineage of nonreturn valves (ball check valve, flapper valve). The "pure" scheme expresses the general working principle of valves, whether mechanical (nonreturn valves) or biological (heart, arterial valves). The artificial heart represents a case of schematic equivalence, of schematic transposability, from the biological to the mechanical fields.

Conclusion

The strength and originality of Simondon's theory of imagination is that it helps us understand this technical genesis and meaningful relation between the subject and its objective world as fundamentally rooted in life, imagination, and technics.

Technical mediations are not mere static tools we use. In Simondon's perspective and in the wake of Merleau-Ponty's insights about cultural integration, technical mediations gain a more anthropologically constitutive role. Life and subjectivity are not reducible to technics. But technics does open up new ways for life to realize and express its own essential powers, in what Merleau-Ponty calls a "third dialectic." Man is, as Merleau-Ponty would say, an integrated living being, a living being extended to, and constituted by, a technical environment he manipulates and creates—and through the mediation of which he imagines and invents.

Simondon provides conceptual tools to help enaction conceive of this "integration" or constitutive articulation between life, subjectivity, and technics. Instead of reducing subjectivity to basic biological and embodied processes so as to make a radical distinction between man and nonliving things such as technical objects—that Simondon calls "easy humanism" (1958, p. 9)—Simondon extends life and subjectivity to the technical world and regards technics as the extension of life. It is not because Simondon thematizes technics that he loses sight of the foundational character of life. On the contrary, thematizing technics is, essentially, thematizing life. Imagination itself is the very process through which living organisms realize their own normativity and extend into their chosen worlds. As I have shown, in Simondon a theory of life is at the same time a theory of imagination and invention. And technologies are nothing else but life, vital, and human motion fixed in the realm of material and informational operative processes. Technical normativity continues vital normativity. Technics does not merely enhance the capacities of organisms. Rather, it embodies their imaginative and active powers. It is in this sense that Havelange, Lenay, and Stewart (2003), for example, think of external representations and technics as ways for organisms to relate to themselves and to achieve their own cognitive functions, such as memory and creativity.

Simondon encourages enactivists to think of imagination as essential to vital individuation and as the very cognitive process through which structural coupling and sense-making are achieved. And instead of taking the relation of coupling between organisms and their technical environments for granted, Simondon's approach allows us to study its imaginative and technical genesis and its very possibility. To say that imagination is technically constituted, in this sense, means that the affordances constitutive of our imaginings are enacted through sensory-motor and technical engagement with the world.

To illustrate it, I have shown how Simondon's notion of practical schematism partially helps answering the question I addressed to theorists of experiential imagination in Chapter 1, which is to know how to think of the intrinsic technicity of instrumentalists' experiential imaginings. I will come back to this question in detail in Chapter 6. But Simondon gives us a general direction that resonates with the pragmatist perspective I offered in Chapter 2, and which I intend to nourish with Ingold's notion of correspondence (Chapter 4), with Stiegler's theory of technical schematism, with Malafouris' understanding of schematism as relational (Chapter 5) and, finally, with ecological psychology's affordance-based account of cognition and imagination (Chapter 6). When imagining playing Beethoven's Waldstein sonata, one pictures himself, in first person, performing with an instrument. But this ability one has, implies that one knows what the instrument affords in terms of expressive possibilities and affective values. One's imaginings have to do with technically constituted possibilities for action. How are we to understand the correspondence between the content of experiential imaginings and the technical milieu in which these imaginings (imagined bodily and instrumental expressions) project us? In Simondon's perspective, the answer lies in the notion of practical schematism, in the idea that image genesis is at the same time a process of technical engagement and concretization. Image genesis and technogenesis are inseparable. Imaginings correspond to a technical context for the reason that they are technically constituted. Imaginings refer to possibilities for actions, to affordances. And these affordances emerge through technical engagement, through a process of technical invention, which Simondon understands in terms of practical schematism, a constant practical and imaginative relation between the agent and its technical milieu. Agents manipulate and transform their technical milieu, so that affordances emerge and individuate, that constitute instrumentalists' imaginaries.

At this point, however, a topical question remains unaddressed. In what sense exactly are images technically constituted? Are they the final product of technical genesis? Or are they technical in themselves, infused and diffused in between interiority and exteriority? Instrumentalists do have private imaginative experiences. These experiences might be enabled through technical engagement, through a kind of practical, technically constituted, schematism through which images co-individuate with the technical milieu. But as private and abstract, imaginative experiences like imagining oneself playing the piano when having no piano in hand, seem, at first glance, be purely internal and representational. After all, the ability I have to engage in such imaginings privately ("mentally" in the traditional, internalist sense of the term) might be acquired through practice, by means of technical mediations. But, again, at first glance, in the very moment the instrumentalist "possesses and manipulates" them, in the intimacy of her closed eyes, these imaginings are neither technical nor external in themselves. Thus, as I announced in the Introduction, the question will be (in Chapters 5, 6, and 7) to know if there is a representational realm of imagination that is irreducible, given its abstraction, to the kind of pragmatist and relationalist perspective I adopt.

4

Materializing imagination

1. Imagination as a Mode of Engagement with the World

The key move, which I unearthed from Simondon's theory of imagination, consists in (1) refuting the commonplace postulate, according to which imagination unleashes us from reality only, and (2) thinking in a nondualistic way the articulation between imagination and (the experience constitutive of) reality. As a reminder, for Simondon, imagination refers to a function of anticipation, perception, and invention. *It is a mode of engagement with the world*, a biological mode of bringing forth an individual-world system that is technically constituted. Images refer to (1) sensory-motor tendencies and spontaneities preparing perception and action (pre-perceptive or anticipative images), (2) modes of apprehension of the object (intra-perceptive images), and (3) "symbols" constitutive of an imaginative universe allowing the insertion of mental activity in the world, in the past as well as in fictional situations. When technologically mediatized, imagination refers to the schematic and participative, ultimately inventive process whereby subjectivity and the technical world co-emerge. *Imagination is the very participative (manipulative, productive and transformative) relation the organism engages with its environment.* And technics plays the role of the material and concrete condition by means of which the organism and its world co-individuate. Let me quote Simondon again:

> Imagination as anticipation is not anymore a function that unleashes from reality and develops in the unreal or the fictive; it initiates an effective activity of realization, for the reason that the subject projecting the image, is the owner of the instruments of production, and the owner of the necessary frame material. (Simondon 1965–1966, p. 56)

Two ideas here: (1) imagination is a mode of engagement with the world and (2) a function of realization in virtue of its intimate relation with mater and technics. Material and technical engagement makes of imagination a function of realization, a mode of bringing forth an individual-world system. These two ideas constitute the two common points of Simondon's and Ingold's perspectives on imagination.

Ingold regrets that we "regard the imagination as an escape from real life rather than its impulse." According to him, the problem of imagination is "not of how to reconcile the dreams of our imagination with patterns in the world, but of how to separate them in the first place." Imagination is not separated from perception. "Imagination" is not just "a word for what does *not* exists" (Ingold 2013b, p. 735). Rather, in Ingold's provocative words, it is a kind of participation to a "more-than-human world" (Ingold 2013b, p. 739) underlying perceptive experience.

Children dream about dragons, cousins of ancient dinosaurs, awaking in terror just before being swallowed by their flaming mouths. In his *Life of St Benedict of Nursia*, Gregory the Great relates the story of a monk who found his path blocked by a frightening dragon. "Convinced that the dragon was about to eat him up, and trembling with fear, he (the monk) shouted to his brothers for help" (Ingold 2013b, p. 736). Are we to say that those dragons do not exist? That being a dream, or a hallucination, is not being real? Drawing from the example of Ojibwa's imaginary Thunder Bird, the pinési, and from Gaston Bachelard's bird of wind and tempest (Ingold 2013b, p. 737–8), Ingold prefers pointing to a kind of *imaginary reality*, to what appears, in Bachelard, as the dreamlike structure of phenomenal experience (Bachelard 1941, 1943).

In Bachelard, indeed, imagination and reality are not exclusive, but inclusive. Reality is not that which we perceive and memorize in order to build fictional images afterwards, by means of fanciful and purely mental combinations. Rather, reality is what we experience in the glory[1] of a conditional, imaginative, and ultimately dreamlike process. Not that, endorsing a caricatural version of Descartes' skeptic argument, reality and dream would be indistinguishable. Rather, a more foundational reality, that of the imaginary, underlies our relation with the world. After all, dreaming of dragons or hallucinating a pinési lying on a rock is an experience. And, as I emphasized in Chapter 1, imagination has to do with the experienceable:

> Thunder Bird makes its presence felt not as an object of the natural world but, more fundamentally, as a phenomenon of experience. (Ingold 2013b, p. 738)

But for Ingold, it is not so much a matter of knowing what kind of experience it is. Rather, the question is to know in which way it is part of experience *in general*. Let us keep in mind Simondon's fundamental postulate: images are intrinsic parts of perceptive experience. They are pre-, intra-, and trans-perceptive. Imagination structures and polarizes experience. Through imagination, the organism virtually explores its future behavior and grasps the meaning of its perceptive or fictional world. In a very close way, Ingold takes experience to be filled with images richer than perception and action. The perception of thunder is part of a richer experience, an experience of imaginative kind. And the boy who, during a thunderstorm, ran out of his tent and saw a legendary pinési, had a "real" experience of thunder—"real" in the strong sense that images filled perception with its very own and intrinsic potentialities. Let us say that, in this perspective, dreams and hallucinations refer to the way imagination actualizes the intrinsic powers or potentialities of perceptive experience itself:

> It (the pinési, or Thunder Bird) is the incarnate form of a sound that reverberates through the atmosphere and overwhelms the consciousness of all who hear it. Just as the monk's brethren, as they rushed outside, saw no dragon, so the boy's parents did not themselves witness *pinési*. But as the conventional shape of a powerful auditory sensation, it would have been entirely familiar to them. The Thunder Bird may be a figment of the imagination, but it is an imagination that has saturated the fullness of phenomenal experience. (Ingold 2013b, p. 738)

Perceptive experience is more than the mere perception of a given physical world. Rather, it is a way to engage with the world. And imagination appears as a primitive, general, and conditional mode of experience:

> The flesh-and-feathers bird is but a manifestation of the real bird of the dream-storm, rather than the other way round, and could not exist without it. (Ingold 2013b, p. 738)

Likewise, the dragon is not just a dream by means of which the monk escapes from reality. Rather, it is the *hyper-real* or synesthetic manifestation of experience's intrinsic powers and lines of force—in this case, an experience of fear:

> The fearsome dragon of Gregory's account was the form of incandescent terror enveloping the subject becoming self-aware at the moment of waking. (Ingold 2013b, p. 738)

Let me summarize this first idea. Imagination is a mode of engagement and the dreamlike beneath of perceptive experience. Images fill perception. We could say that beyond sensoriality understood as the mere reception of sensorial information (information coming from the environment, as well as from one's own body), sensitivity refers to the affective and imaginative coloration of perceptive experience. Images reveal the intrinsic affective and aesthetical values, that is, in one word, the meaning of perceptive experience.

This view on the foundational status of imagination applies to action as well. A good illustration of this is musical and instrumental performance. Playing the cello is not just producing sounds with a cello. It is not just having a sensorial and kinaesthetic experience of a sounding cello. Rather, it is, in the very moment I am playing the notes written on the score, grasping and expressing the intrinsic and imaginative powers constitutive of the sounding relation I engage with the cello, the venue, and the audience. Playing the prelude of the fifth suite (BWV 1011) is not just making the C and the G strings vibrating. It is engaging with, and expressing specific, imaginative lines of force. The C and the G strings resonate together like a powerful rumble coming from the bowels of earth, merging in a hallucinatory but serene reflection of autumnal daylight. In a different register, playing the Corrente of the second suite (BWV 1067) does not merely consist in playing alternatively on the D and the A strings only. Rather, following Ingold's image, it might consist in becoming some kind of pinési. But the living pinési here must be understood as the revealed imaginary beneath of Bach's musical text. Thunder triggers an imaginative world, gives birth to legendary birds. The same goes with the cello. The strings, as well as my bodily engagement in the bowing and the fingering, trigger imaginative and expressive lines of force. Put simply, if playing the cello consists both in performing and in perceiving sounds, imagination appears as the very dynamics whereby I engage with the cello, the sound, and the environment.

Worth noting in the light of this description, Ingold extends Bachelard's materialism about imagination. Bachelard defines imagination as being "material." The idea is that the forms of imagination are isomorphic to those of material elements, air, water, fire, and earth. Material elements are vehicles of images. Regardless of its singular forms, matter has its richness and depth, and exhibits affective qualities. Earth suggests different poetical images and emotional qualities, like those of stability and protection (the asylum) or anxiety and perdition (the labyrinth, the deep pit). Air supports dreams of flight, of birds and dragons. As Ingold puts it, quoting Bachelard, "the bird 'is the dynamic eye of the storm' (Bachelard 1988, p. 77): its body the wind, its breath the tempest, and its wings the sky" (Ingold 2013b, p. 738). Matter penetrates images. Images have with material elements an isomorphic and dynamical relation, exhibiting specific affective and emotional values.

As a material anthropologist, Ingold interestingly extends this materialism about imagination to material culture. The instrument fills the cellist's imagination with a specific set of forms and affective values. Coming back to the example of the BWV1011 prelude mentioned above, very concretely the A string, detuned in G, as Bach himself recommended, vibrates more slowly, more weakly than usual. Its vibration amplitude approaches the amplitude of the lower D string. Both G (A detuned) and D strings merge in a larger and somehow homogeneous sound. The sound is less penetrating, more resonant, thicker, and more evanescent at the same time, giving this deletion and fateful effect. Here, what appears clearly is that *the very form and affective value of the musical image, lies in the instrument setting and manipulation*, in its material and vibrating behavior. The image is not something I could imagine by myself, independently from the cello, before I play on it. Rather, the image is given to me by the instrument. Detuning the A string in G changes everything, surprises my imagination, opening up a totally new world of affective and expressive engagement. In this sense, playing the cello is being driven by these imaginative and material engaging powers into the realm of plain sensitivity, beyond mere sensoriality. Let us consider the mere difference between playing on gut strings and playing on metal strings. The difference matters to such an extent that they correspond to different repertoires, to different styles, technics, schools, and, finally, to different musicians. All goes as if imagination depended on the material objects we engage with, in accordance with its isomorphic relation with the forms and affective values exhibited by those objects.

2. Ingold's Approach to the Technical Constitution of Images

2.1. Baroque Bowing and the Material Nature of Musical Images

Instead of pressuring my bow on the string, I just use the weight of my arm. I let my arm fall onto the string. Pressuring implies a tiring muscular exertion that does not produce the deep sound, and the smooth and volatile articulations I am looking for. The rosin on the hair bow increases the adhesion to the string. The wood of the bow, of the bridge, and of the sounding board of my cello, and my own fingers, hand, wrist, elbow, and shoulder, bend and resist at the same time. The proportion of these bending and resisting, the weight of my arm in the bending-resisting taut string, determine the sound, its density, color, and power, between violence and fading, plenitude and uncertainty, resonance and penetration. The resistance phenomenon is a striking and omnipresent one in instrumental practice. Musical articulations and the instrumentalist's touch, whether it is on keyboard, bowed or plucked string, or wind instruments, depend on this proportion of resistances. The resistance is proportional to the strings' tension. And the way and the time the sound lives, lies in the real-time management of these proportions.

But these proportions change according to the concrete instrumental device I choose. Let us say that I have the choice between four different bows and two different cellos. Two of my bows are made in baroque style, one is classical, and the fourth is a modern one. The baroque ones have a concave shape. The first one is a Cangelosi, made in Florence in 2002 according to a eighteenth-century Italian model. It is a light, reactive bow, Snakewood for a soft, rich, and nervous sonority. The second one is close to a viola da gamba bow. Longer than the Cangelosi, this baroque bow is made in a heavier Pernambuco wood. It provides a more powerful and homogeneous sound, but less-precise and less-sharpened articulations. The sound is like merging. The classical bow is convex and light at its tip. Classical repertoire, sharp articulations, the sound is less resonant, more penetrating. The modern bow is convex the same. As well as the classical one, and unlike the two baroque ones, the modern bow makes it possible to accentuate the bowing attack at the tip of the stick, with almost the same power and clearness as on to the frog. Heavier at the tip than the classical bow, the modern helps compensating the loose of weight in outstretched arm. The sound is sharper and more powerful than with any other bow, more penetrating than resonant. The baroque bows are particularly well suited to little spaces and little audiences, while the modern one is adapted to large spaces and audiences. With the Cangelosi, I can play Vivaldi, Boismortier, Barrière, or even Boccherini's sonatas for solo cello and continuo (a second cello and a harpsichord). The modern bow gives me the possibility, if used properly, to penetrate the sound space in such a powerful way that I may play the solo part of a cello concerto, over a seventy musicians orchestra. The question I ask myself is the following: "are my neighbors at home?" More seriously, "which musical repertoire and musical intentions am I concerned with?"

All does not rely on the bowing only, for sure. The way my cello is made is a highly determining factor too. Without dwelling on it here, the nature of my musical images differs whether I play on a Franco-German modern cello, industrially made in the 1990s, high fingerboard inclination, Belgian bridge and metal strings attuned in A 440Hz (maximum string stretching and sound penetrating power), or on a re-historicized 1730 Austrian cello, low fingerboard inclination, gut strings attuned in A 415 Hz (for a more resonant sound). From an instrumental device to another, utterly different musical and imaginative worlds open their doors. What I imagine, as a musician, the value of my musical images, changes with the instrument. The instrument provides its own musicality, determines my own musical imagination. Not that I need a baroque cello in order to play Bach instead of contemporary composers like Kodaly or O'Connor, nor that playing Bach with a modern cello is absurd. Rather, given the instrument making, I will forge and express different kinds of musical intentions while playing these musical pieces. As Gadamer would say, Bach's musical texts are full of promises to be fulfilled. But I will add that those promises do not come from the text only, but also and even essentially from the instrumental practice and making.

Say that I choose the Austrian baroque cello with the Cangelosi bow. I bow up on the first note, string switching, and left-hand position changing for the prelude of the fourth suite in E-flat Major (BWV1010). I will bow down in anacrusis, at the beginning of the following Allemande. Each bowing is—as well as the time I take to change my left-hand position and then to reach the following note, as well as the time I need to switch from the C string to the A string—and each of my technical move is a gesture, that is, a move through which I relate to my cello and through which, in the very same time, I give birth to musical images. Bowing consists in engaging my whole body, from the ground to which I relate all the more so as I play with a baroque cello (no endpin) to my fingertips. And the way I bodily engage in the string is itself the realization of a musical image.

But what does the cello bring into the equation? My claim is that the variety of emotional experiences and the breadth of the affective and expressive spectrum provided by such or such instrument (whether it is a baroque cello, an Oud or an electric guitar) give the musical and instrumental interpretation its richness, its finesse, its ambiguities, and its clarities. It provides to musical imagination a particular and constitutive set of emotive musical behaviors. The instrument shapes my musical imagination, in the very moment I am playing on it. I bow up. How will the string behave? What will the outcome of my move with the cello be? Exactly, I do not know. I can feel it. I have a nonconceptual knowledge of it. I can anticipate the sound in its virtual generality, but not in its concrete singularity. A kind of instrumental and auditive imagery through which I engage with my instrument directs me. I am looking for something; I tend to produce such or such sound. I have a hunch, sort of a musical-bodily instrumental impression I forged through practicing, through instrumentally situated and embodied experiences. But this impression, anticipative or, in Simondon's terms, pre-perceptive, about to become musical performance (intra-perceptive), is not in a move which I apply accidentally to such or such instrument. My move is that move and not another one, with a high and imperceptible degree of precision, given the instrument to which it applies. When she plays, or imagines playing Beethoven's Waldstein sonata, does the pianist play on a very sharp, with absolute brilliant clarity Steinway, or on a Bösendorfer, with a warmer and richer tone, remarkably well-structured? Heavier key actions can also create a depth of tone and much richer sound than a light key piano. What does it mean as regards the kind of rhythmic, affective, and expressive values the pianist will forge, whether concretely or in imagination?

The cello and the piano behave and respond according to their own material and technical logic. Instrument manipulation has its intrinsic and unforeseen accidents. The instrument has its own weakenings, constraints, and colors. Bach would talk about the unique grain of the instrument, which makes the very soul of a given interpretation, and thanks to which the musical image differs from pure auditory, disembodied, and dematerialized information. This accidentality, this unexpected instrumental variability and the way it determines musical interpretative imagination are the very mark and destiny of baroque musicians. The baroque instrument essentially differs from the modern one due to its higher degree of accidentality. String tension decreasing increases accidentality and unexpectedness. The baroque instrument is more sensitive and less docile, more capricious, and uncertain than the modern one. The gut string and the metal string offer two different sets of practice conditions. They determine different imaginative, gestural, and imaginative attitudes and behaviors. The gut string is sensitive to humidity (to sweating), to dryness, and to temperature, and detunes in the middle of the performance. Never the same and always determining, the gut string illustrates what this instrumental power is and the irreducibility of the imagination to pure and voluntary mentality. There is a variability and accidentality that is intrinsic to material things, let us say, a material life of things that counts as the very living form of the musical image.

What do these descriptions mean as regards the technical constitution of imagination? How are we to understand that the cello, or the piano, is more than external means for internal imaginative processes?

A "causal coupling" explanation would claim that musical imagination is purely representational and that the embodied use of external and instrumental devices gives rise to imaginative representational processes without being imaginative per se. The cognitive and imaginative system may be coupled to external technical devices, bow and cello in this case. However, coupling does not imply ontological confusion. Neither gestural nor instrumental *constitution* of imagination (ontological identity). Just *causal coupling* (external causation of internal imaginative and representational processes). Accordingly, from the point of view of representationalists, we can definitely say that the sound we hear in our minds, constitutive of our musical images, is the sound of my cello. Mozart, who appreciated composing for singers he befriended with, knew their voices and creatively combined auditory images of their voices. Musical images, as mere auditory images just root in my perceiving the sound of my cello or of anybody's voice. The auditory image, constitutive of my musical imagery, is not my voice or my bowing in themselves, just the internal and mental quasi-perceptive rehearsal of the sound they produce. Whether we endorse Clark's parity principle (Clark and Chalmers 1998; Clark 2010), according to which the technical device is cognitive because it is involved in the realization of a cognitive function, or Adams and Aizawa's point of view, according to which *functional realization* does not mean *ontological identity* (2008, 2010; see also Rupert 2004, 2006, 2008), my musical images in both theoretical perspectives remain, in themselves, representational and internal processes, devoid of any external-instrumental aspects. Another example, the organist chooses her registration. But the registration technics is not in itself imaginative. It just gives rise to an imaginative world of sounds and colors, that the composer or the organist mentally combines (in order) to create their own musical work or instrumental performance. In the same way, the composer chooses her orchestration without playing each instrument.

But I cannot take the idea that musical images equal auditory images, eventually combined with linguistic-like structures in a "musical language," for anything else than a perfect misunderstanding of what music essentially is. This reductive perspective ignores what instrumentalists, singers, and composers still know as fundamental evidence, namely that the question is not just about sound but also about bodily, instrumental, and vocal singularities, capacities, and skills. Composers do not compose for *the sound of a voice*, but for *a voice*, understand, a set of sensori-motor patterns, of bodily attitudes, gestures, qualities, efforts, abilities to place, articulate and perform sounds, textures, emotions, and so on. They do not compose for the sound of the piano. Rather, they compose, in the best case (1) for a set of technical skills and capacities thanks to which it is only possible to play such or such passage on the piano and (2) more generally for a specific kind of material, both bodily and instrumental life, proper to such or such instrumental and cultural practice. The idea I defend here might be provocative and counterintuitive. And I shall justify it in the language of cognitive science in both Chapters 5 and 6. For sure, it is clearly not self-explanatory: musical imagination is not just about sound but more fundamentally about how we perform a musical work, that is, how we behave through it, when dancing, performing, singing, mimicking, describing it. When listening to a musical piece, I do not hear a mere combination of sounds. Rather, I participate into the material life of a behaving and sounding world. I access a material behavior, that of the instrument, thanks to its sounding resonance, and that of my body when I engage in instrumental practice. My images are material in the sense that their constitutive forms and affective values are those exhibited by mater. The sound is just a perceptible effect and reflection of this behavior. But the sound appears as nothing more than the surface of the musical image.

2.2. Ingold's Approach to Material Imagination

Ingold straddles images and material objects, situates mentality and imagination on the borderline between internal and external processes. He identifies the flow of consciousness with the flow of materiality. If we momentarily stop and conceptually make a distinction between both flows, we get on the conscious side, a given fugitive image, and on the material side, a given state in the growing materiality, what we usually call an object. But to say that both flows are one and the same (Ingold 2013a, p. 20) is to say that there is no way to understand the genesis of images, that is, properly said, "imagination," independently from the materials and genetic life of things. The genesis is not that of mental images afterwards imposed on the material world. Nor is it the genesis of matter only, afterward integrated under the form of mental representation in the representation box of the agent. Rather, imagination is a relational and co-participative process at the end of which only images and material objects co-emerge.

What is at stake for Ingold is the characterization of this object-image identity in a nonhylomorphic way. Hylomorphism says that the making of artifacts, or of a work of art for example, consists in imposing a form internal to the mind, upon material given things. This is a dualist view. The mind appears as outside of the world, with its own mental images or representational forms, and the world as the undetermined and non-determining support of mental projection and realization. Mind and world face each other in a domination kind of relationship, where the imaginative agent imposes his mental force on the object. On the contrary, and explicitly following Simondon (Ingold 2013a, p. 25), Ingold aims at thinking a process of material and imaginative *growth*, in terms of joint forces, of what Simondon calls *morphogenesis*:

> To read the making longitudinally, as a confluence of forces and materials, rather than laterally, as a transposition from image to object, is to regard it as such a form-generating—or morphogenetic—process. (Ingold 2013a, p. 22)

As I explained earlier, the criticism of Hylomorphism as a way to think morphogenesis is the very starting point of Simondon's philosophy of individuation. As a reminder, Hylomorphism comes from Aristotle and consists in explaining genesis in terms of matter (*hyle*) and form (*morphe*). Matter and form preexist to their union. Matter takes form. The preexisting form is applied to the homogeneous, undetermined, and nondetermining matter. As a reminder, Simondon shows that the hylomorphic schema is insufficient when it comes to explaining true genesis. For the individual form is what is to be explained as the result of genetic processes. And to explain genesis as the application of a preexisting form to neutral matter is a cyclic argument. Again, what we want to explain is presupposed prior to its "explanation": the individual is presupposed prior to its individuation (genesis).

In Simondon's and Ingold's perspectives, the ontogenetic principle of morphogenesis is not the form, but matter itself. In line with Gaston Bachelard's material existentialism (common denominator of Simondon and Ingold's philosophies), Simondon situates the genetic operating processes in matter. As to Ingold, he understands imagination as a material and relational process involving material culture. Genesis is a material process. And materials are not, in the words of Karen Barad quoted by Ingold, "'little bits of nature', awaiting the mark of external force like culture or history for their completion" (Ingold 2013a, p. 31). Instead, they are the "substances-in-becoming" (*Ibid.*), and they are the dynamical substance of a world in formation. And as would say Spinoza, Man is not, in Nature, like an empire in an empire. There is no metaphysical difference between imagination and material-natural growing. Imagination and creativity occur as constitutive parts of the material "life of things." Comparing—as Simondon would do with his analogy between crystallization and individuation—stalagmite formation with marble sculpture, and, like Simondon, bringing "things back to life" (Ingold 2013a, p. 20), Ingold defines imagination and creativity as a mode of material growing (morphogenesis):

> This is to soften any distinction we might draw between organism and artefact. For if organisms grow, so too do artefacts. And if artefacts are made, so too are organisms. What varies, among countless other things, is the extent of human involvement in the generation of form: but this variation is one of degree, not kind. (Ingold 2013a, p. 20)

According to Carl Wieman, there are two ways to define creativity (see Ingold 2014, p. 126). The first refers to creativity as proper to the capacity a human being has to construct "something according to a new design that has already come within reach of his imagination" (Wieman 1961, p. 65). Accordingly, imagination consists in the mental previsualization of images understood as mental forms or designs. This kind of creativity is *done*, for it depends on the pure voluntary mentality. The core naïve idea of this hylomorphic view is that, in my mind, I imagine something and realize it in the world afterward. Such an understanding is, I might say, devoid of any reality principle.

Ingold rejects this explanation. He defines imagination in terms of an *undergone* (rather than done) genetic process. I do not play on the cello whatever I want or imagine. I play only and especially what it is possible to play and to imagine with *that* cello. And the singular affordances of my 1730 Austrian cello and Cangelosi bow are different from those of my 1990's Franco-German cello and modern bow. Material culture is then part of the imaginative process.

"Fine" one might say. But the maker or the instrumentalist, would say an internalist, "may have an idea in mind of what he wants to make. He may even be seeking to copy a piece of work that already stands before him. Does this not distinguish the statue from the stalagmite, once and for all? Can we not speak, in a sense unique to artifacts, of their *design*?"

Here is how Ingold answers to this obvious objection:

> If imagination is not about the composition of novel designs in advance of their execution, as a condition of *doing* in Wieman's sense, or about the origination of "creative ideas" in Boden's sense of P- or H-creativity, then what is it? . . . I want to argue that imagination is not a mental capacity that permits the spontaneous generation of ideas, but rather a way of living creatively in a world that is itself crescent, always in formation. To imagine, as anthropologist Stuart McLean puts it, is "to respond creatively to the creativity of the world's ceaseless self-transformation" (McLean 2009, p. 231). This correspondence—this answering to a world that, in its relations and processes, also answers to us—is the generative dynamic that moves life forward, and which leads by aspiration. (Ingold 2014, p. 134)

Creativity also refers to what a person undergoes but cannot do (Wieman 1961, pp. 65–6). This kind of creativity does not begin here with an idea and end here with an object. "Rather, it carries on through, without beginning or end" (Ingold 2014, p. 127). According to Wieman, this kind of creativity suits for the explanation of social phenomena. It is proper to social life, to the progressive creation of personality in community, for example. Wieman is a reader of Whitehead and, through him, of Bergson, two philosophers who thematize life as their central philosophical object. In Wieman, Ingold appreciates Whitehead and, through Whitehead, Bergson. He is sensitive to Bergson's insights about the genetic essence of things (see Bergson 1889, 1896) and about the social roots of individual personality. Ingold takes from Whitehead the concept of *concrescence*, which refers, as a direct and explicit tribute to Bergson's "*durée*," to the "capacity of living, growing things continually to surpass themselves" (Ingold 2014, p. 128). What is created does not exist as a static thing in a geometrical space. Rather, it is part of the world, of a world-in-formation. In one word, it is part of *ontogenesis*. And to say that an object is part of ontogenesis means, in Ingold's perspective, that it carries its own genetic powers, in the same way living things have their own "*durée*," that is, their own principle of concrescence.[2]

As I emphasized previously, a crucial aspect of Simondon's philosophy of technics is precisely to recognize to technical objects their own mode of existence. Through the concept of "auto-correlation," he makes a constant analogy between livings and technical objects. Like the former, the latter have their own concretizing genesis, their own ontogenetic principles.

In Ingold's perspective, this genetic essence of materials characterizes material culture as well as imaginative and creative processes. "What Bach has done, Ingold says, was rather to launch the music into the world" (Ingold 2014, p. 239). Bach launched the music into our imagination, as a worldly salient existing work of art. And century after century, we continue to be part of its concretizing genesis. But the equation does not combine the internal mental imagination with the external material score. Rather, it synthesizes the two in a new and non-hylomorphic way of understanding imagination as a *material revealing process*. Ingold refers to an interview with Patricia Cain, in which sculptor and draughtsman Richard Talbot says, "I don't think I think with images. But the drawings . . . when I'm setting out to the drawing, I don't have a pre-conceived image. . . . I might have a hunch" (Ingold 2013a, p. 126–7). In the same way and as I said, when I play the cello, I do not anticipate in its singularity the outcome of my bowing, as if my musical images were pregiven in my own and isolated mind. My musical image engages me with the feeling of what I intend to do, without it being done. Rather, it is undergone, in the sense that the concrete bowing, the concrete manipulation of my instrumental device feeds and directs the genetic, the so called imaginative process. I follow-direct my cello. And my musical image reveals in the bowing. I literally catch the image, under the form of a conscious (partly unconscious too), bodily and instrumental experience, in the very moment it lives in the material world.

As Ingold duly notices, this description tallies with what graphic artists have to say about their work, namely, that the essence of drawing as an imaginative activity, does not lie "in the projection onto the page of interior mental pictures." Rather, aesthetical lines reveal themselves in the core of the material and technical engagement with tools and instruments. Imagination is in the instrumental manipulation, as both a revealing and what Ingold calls a *corresponding* process. Correspondence, "this answering to a world that, in its relations and processes, also answers to us," is a good description of what playing the cello consists in. For sure, bowing is not being mastered by the instrument. It is not being absolutely passive. And even more, it is engaging actively in the string. But the difference between the specialized performer and the non-cellist lies in that the specialist knows how to listen to her cello's behavior and how to answer it, without forcing it to behave in an arbitrarily expected way. The good cellist feels the way her cello behaves and what to do in order to get a given behavior from it. But once again, musical images are not mental forms one applies afterward to the cello. The cello's behavior (sounding, vibrating, the way the bow goes on the string, etc.) gives an ontic part of the image in real time. The musical image is in how the cello behaves when the cellist behaves this or this way. It is not just a mental representation of it. The cello's behavior and the musical value of this behavior, which, with the cellist's behaving, constitutes part of the musical image, are not represented in the mind, preexisting to the concrete manipulation. It is revealed in the core of the instrumental manipulation. The image occurs as the product of a technical system coupling the cellist and the cello. And the very singularity of the bowing, that is, of the revealed musical image itself, is the outcome of a real-time singularization process that consists in bowing this or this way. The sound's singularization consists in bowing imaginatively, that is, in listening and answering, directing, and being directed, in one word, *corresponding*:

> From here it is but a short step to the conclusion that drawing that tells is a correspondence, of kinaesthetic awareness and the line of flight. In this correspondence, as Bryson says (2003, p. 154), the "mark on paper leads as much as it is led", alternately sewing the line into the mind and the mind into the line in a suturing action that grows ever tighter as the drawing proceeds. Thus the drawing is not the visible shadow of a mental event; it is a process of thinking, not a projection of a thought . . .. Instead of dictating a thought, writes Pallasmaa,[3] "the thinking process turns into an act of waiting, listening, collaboration and dialogue [in which] one gradually learns the skill of co-operating with one's

> own work." Co-operating with one's work—now there's a good definition of correspondence! This thinking, this imagining, goes on as much in the hands and fingers as in the head. (Ingold 2013a, p. 128)

I build up my musical images following the flow of materiality. And following is precisely what differentiates imagination as morphogenesis, from imagination as hylomorphism. As says Ingold, citing Deleuze and Guattari reassuming Simondon:

> The trouble with the matter-form model . . . is that in assuming "a fixed form and a matter deemed homogeneous" it fails to acknowledge, on the one hand, the variability of matter—its tensions and elasticities, lines of flow and resistances—and, on the other hand, the conformation and deformations to which these modulations give rise. In reality, they insist, whenever we encounter matter "it is matter in movement, in flux, in variation," with the consequence that "this matter-flow can only be *followed*." (Ingold 2013a, p. 25)

Correspondence is then the core operation constitutive of material and instrumental imagination. A cellist corresponds with his instrument. Otherwise, he would not be a cellist. He just might impersonate the cellist in general, like non instrumentalist actors do. This correspondence is a specific kind of *interacting*. Ingold describes it in the lights of Alfred Schütz insights about social and especially musical cases of interaction (Ingold 2013a, p. 106). In Jazz improvisation, the players do not exchange musical ideas that they would translate afterwards into bodily moves applied to the instrument. Rather, as Schutz puts it, they "move along together, listening as they play, and playing as they listen, at every moment sharing in each other's "vivid present' (Schutz 1951)." Musicians imagine in real time, feeling each other's embodied and imaginative presence, forging a collective and embodied imaginative ongoing process. Musicians imagine together, corresponding to each other's behavior. The personality of a musical group is socially created, through this undergone and relational kind of creative imagination. The same goes for the individual personality of a given cellist, whose personality, as a cellist, progressively rises through her concrete practice, through her correspondence with others and with her cello.

Against hylomorphism again, Ingold describes this correspondence-imaginative process by using a term central to Simondon's philosophy, which I already defined in the previous chapter, the term "transduction." Again, a transductive process is the simultaneous individuation of inseparable domains of reality. Then, a transductive imaginative system is not accountable in terms of a coupling between preexisting images and preexisting materials (the pencil and the line). Rather, images and artifacts are the a posteriori outcome of a common operation where "mental" means both sensory-motor-emotive engagement and material manipulation as well. Through action-perception, imagination takes place on the borderline between the mind and the line, between interiority and exteriority. Mind and line, interiority and externality only result from this previous genetic and indivisible process. As I emphasized, my musical image is not prior to instrumental manipulation (as would say the hylomorphic view). Rather, it appears as both a sensory-motor-affective *experience*, and a material-instrumental *performance*. Ingold refers precisely to this ontogenetical identity proper to material imagination, by citing English writer and painter John Berger's formula: "drawing-thinking" (Ingold 2013a, p. 128). Malafouris would talk about a process of "thinging" (see next chapter). Thinking is drawing, and drawing at the same time is in itself a thinking process.

Finally, as transductive, the imaginative system reaches states that are not deducible from an analysis of its previous states. Ingold remarks with Bergson that we cannot read creativity backwards, from the musical work or the instrumental performance for instance, to the intention that motivated it. The intention, in its richness, is more than what an isolated mind could compute by itself. It is *improvised*, that is, it is the outcome of a relational process implying more than my own computing mind. It implies the world-in-formation's singularities. It surprises me as the result of something more than what I already am, as an isolated imaginative agent. Put in another way, there is, in the image, something unexpected that comes directly and only from the instrument and from the material world, from their "material variations."

Thus, a transductive system is a relational and dynamical system, of which morphogenesis does not imply the preexistence of separated forms and materials. It is an ongoing creative process. Concretely, in the case of bowing, this transductive interaction consists in following-being-followed by the bow. The bow, as well as a pencil in my hand, is a *transducer*, which "converts the kinetic quality of the gesture—its *ductus*—from the register of bodily movement and awareness to that of material flux" (Ingold 2013a, p. 128). As a transductive process, my musical-instrumental imagination does not involve prior images separated from a given instrumental device. Rather, my musical image, as well as the instrumental expression of it, is the outcome of a same prior genetic process, instrumental, material, and mental. And this prior ontogenetic process is the "transducing," the "correspondence" in actuality. The transducer makes the ontological and aesthetical equivalence between my bodily move and my cello's sound, between my behavior and my cello's (material and sounding) behavior. As such, the transducer, the Cangelosi bow in my case, is not, as put by Pallasmaa, "a bridge between the imagining mind and the image that appears" in the resonating sound.[4] It is not, neither, the mere means of new internal imaginative experiences. Rather, it is, in Simondon's words, the relational center of genesis—in this case, of imagination understood as the conscious, material, and instrumental genesis of images. Put another way, the transducer (the bow or the piano key, for example) is a mode of participation to the material life of things. It is a way to musically work with mater, to musically behave as part of the material life of things.

Conclusion

Let us take stock on the arguments and perspectives—still philosophical and in need of clarification from the point of view of cognitive science—all this provides on imagination and its technical constitutivity. First, in order to think this constitutive relation between imagination and technics, we need not to reduce imagination to a mere faculty of the unreal. Instead, imagination appears in Simondon and Ingold as a general and conditional mode of engagement with the world, where technics plays a central role.

In Ingold more specifically, imagination is the source and dreamlike beneath of perceptive experience. This is a significant philosophical gesture. Imagination is not separated from perception and action. I enlightened this through the analysis of instrumental performance. Instrumental performance constitutes a particularly good illustration in that it is a paradigmatic kind of engagement with the world, where both perception and action reveal their intimate relation to each other, as well as their intimate relation to imagination and affectivity. In the wake of what I suggested in Chapter 3, especially in the light of Simondon's theory of the image cycle, this leads to a strong idea: both perception and action root in *imagination understood as the general and biological mode of perceptive and agentive engagement with the world*. This, as I will show in Chapters 5 and 6, appears as a way to articulate material culture to life.

Second, images exhibit forms and affective qualities that are to be found directly in the texture of external and material things. There is an isomorphic relation, a kind of formal identity between images and material things. As a reminder, Simondon thematizes the same kind of isomorphic relation between the schemes of imagination and the concrete structure of external things.

Imagination incorporates the operative modes that are constitutive of natural and non-natural, as well as of human and nonhuman things (Simondon 1961/2006). However, Ingold's theory of material imagination combines two particularities. First, it extends this materialism about imagination to material culture. By doing so, Ingold enlightens an intimate relation between the forms of imaginations and the material life of cultural objects. Second, Ingold explains this isomorphic relation in terms of a technical and material genesis of images that help thinking of musical and instrumental engagement.

At this stage of the reflection, there are two complementary ways to define imagination. First, there is imagination as a conditional mode of engagement. In Simondon, and in a sense very close to the enactive approach (see Chapter 3), imagination is a biological mode of engagement, whereby the living organism and the world co-individuate. In Ingold, imagination is a primitive, general, and conditional mode of experience. To quote him again, perception "could not exist without" imagination. Perception is a singular "manifestation" of imaginative lines of force that take root directly in the texture of reality (Ingold 2013b, p. 738).

Second, imagination thus understood, is material for two reasons: (1) its constitutive forms are isomorphic to the intrinsic forms of material things; and (2) the genesis of images is material, consists in a technical engagement that makes us participate in the material life of things. More specifically, imagination is material in two ways. First, imagination is material in a "natural" sense. Borrowing from Bachelard's notion of "material imagination," Ingold takes for granted that the forms constitutive of images are isomorphic to the four material elements. As I just emphasized Simondon's view is quite similar. Second, Ingold extends this materialism about imagination to material culture. Then, imagination is material in a "cultural" sense. The forms of the cellist's musical imagination are isomorphic to the concrete forms exhibited by the cello's behavior. And those imaginative forms differ whether we talk about a seventeenth-century cellist playing on gut strings, about a twentieth-century cellist playing on metal strings, or about a contemporary electronic music composer using any kind of device reproducing the sound of a cello on a midi clavier.

To conclude, technical devices, as transducers, mediate the very genesis of images understood in terms of technical and affective engagement with the material world. And to understand why a relation of constitution differs from a relation of realization, entails abandoning the instrumentalist conception of technics, as well as the hylomorphic paradigm which it carries.

5

Anchoring Imagination

Placing the enactive approach in a mutually informative relationship with Simondon in Chapter 3, I formulated one corollary hypothesis, which is that in order to think about the technical constitutivity of imagination, one needs to conceive of imagination in its essential relation to action and invention. In an enactive way, I defined imagination as a mode of sensory-motor engagement, a biological mode of bringing forth an individual-world system. Drawing from Ingold's perspective, I highlighted the material nature of images.

In this chapter, I will defend two ideas. First, I will show that imagination is not a timeless and world-separated faculty to produce novelty, but rather a disposition to produce images that renews itself depending on the history of the material and technical environment. Imagination shares with technics a same history, differs from a technical context to another, in terms of procedures and operative modes. In this perspective directly inspired by Bernard Stiegler's organology (Section 1) and Malafouris' cognitive archeology (Section 2), it is possible to study imagination by directly studying the technical devices whereby material images appeared in the world. In this sense, acknowledging the technical constitutivity of imagination is emphasizing, as I said in introduction, (1) its technical relativity and (2) its irreducibility to purely internal and representational processes, isolated in time and from the technical environment.

Second, I will show that imagination taken as a material and technical process of image genesis, beyond representing the world, brings forth new ways of perceiving it, of acting within it, and of thinking about it. In Chapters 3 and 4, I emphasized that imagination structures our perceptive and agentive experience. Imagination appeared as a process of sensory-motor engagement. And it is an emotive, inventive, and, ultimately, productive one. Here I want to be more specific. I will argue that the technical production of material images is *itself* part and condition of cognition, shaping the way we perceive, act, and think. It is not just saying, like in Chapter 3, that perception and action are filled with sensory and motor images richer than perception and action, and that image genesis is a corollary of technical schematism and concretization. It is not just making the observation I made in Chapter 4, that our perceptive and agentive experience roots in a dreamlike world, expressing its intrinsic values and qualities. Rather, it is saying that imagination, as such a technical and material engagement process, *enables us cognitively speaking*, transforms the operative modes constitutive of perception, action, and thought. I will define imagination as a technical and material process of cognitive enablement, enlargement, and shaping. This, as I will emphasize, leads to explain the imaginative schematism in extended and non-representationalist terms.

1. Stiegler and the Technical Essence of Imagination

In this chapter, I will develop my reflection in line with what is called the "technics as anthropologically constitutive" thesis (the TAC thesis), proper to the school of Compiègne, of which Stiegler was a founding member. As presented in an enlightening text by Havelange, Lenay, and Stewart (2003), the core idea of the TAC thesis is that technics makes human intelligence possible and shapes it. Tools, instruments, technologies, information and communication systems, softwares, interfaces, and material and political organizations shape the way we think, perceive, memorize, reason, create, decide, imagine, and so on. They structure the ways we define values, desires, identities, and belongings, as well as the ways we meet and interact with people, and the ways we behave individually and collectively (see Pierre Steiner 2010, p. 7). They literally constitute and structure the dialectical relation whereby organisms make sense of their world. They shape and diversify their cognitive abilities. In this perspective, human intelligence is essentially artificial, in the sense that the living organism, with its affective body, brain, and intrinsic dynamisms, does not possess by itself the ability to count, to anticipate, to imagine, or to communicate, for example. These abilities are made possible and diversified through time and history, due to technical resources and changes. Technical objects shape future acts of counting, anticipation, imagination, or communication (see Havelange, Lenay and Stewart 2002). Human intelligence changes along with the history of its constitutive technical mediations.

But before anything, let me insist on one crucial point. This may help understand what I intend to say in this chapter, and prevent any misunderstanding about what I take to be the ontological status of technics with regard to imagination and cognition in general. As I previously emphasized, a philosophical project aiming at understanding the conditional character of technics with regard to human cognition urges us to adopt a noninstrumentalist view on technics (Stiegler 1994). Technics is not just a means to an end. A relation of constitution, as I formulated it, has nothing to do with a relation of realization (see Chapter 4). Technics is not just a way to adapt, according to an already-given order of ends. Ends are not to be found in the intrinsic structure of a given representational mind; neither are they to be found in a world of objects that would be meaningful independently from the organism's activity; nor are they to be found in the organism taken independently from its participation in technical concretization (Chapter 3). Rather, ends emerge in the dialectical and manipulative relation the organism engages with the world through technical mediations. To put it another way, if we agree with the TAC thesis, ends do not exist in the form of a pregiven set of mental representations about the world and ourselves, representations according to which technical objects would be made and used. Instead, they emerge through objects making and manipulation, as a cognitive, social, and technical process of individuation.

However, and reversely, acknowledging this constitutive role of technics is not to say that technics is the only and sufficient condition of humanization and cognitive enablement. In this sense, as I intend to show in this chapter, a relation of constitution has also nothing to do with a relation of reduction. The central notion here, that of "originary prosthetic enablement" (*habilitation originairement prothétique*; Steiner 2010; see below), means that technics enables us cognitively speaking. But this does not mean that technics has by itself the power to determine or predetermine us to such or such cognitive or social behavior. "Constitutivity" means condition of possibility, not causal determination. And as put by Pierre Steiner (2010, p. 26), technics "does not constitute *the anthrôpos alone*." Let me put things clearly, the argument just consists, against the traditional and devaluing instrumentalist conception of technics, in rehabilitating technics as an explanatory resource in the understanding of cognitive, cultural, and historical phenomena. Technics is no more, but no less important than biological, social, political, economic, and religious factors in the explanation of human intelligence.

Now, to go back to the TAC thesis as developed in the field of cognitive science by Havelange (Havelange, Lenay and Stewart 2003; Havelange 2010), Lenay (2008), and Stewart (2010), it roots in the works of André Leroi-Gourhan, Gilbert Simondon, and Jacques Derrida.

Stiegler is the one who first synthesized the perspectives of these authors, integrating their achievements in a powerful reflection about the technical constitution of subjectivity. And he does so by developing an original theory of the imaginative schematism. Stiegler's thesis about imagination might be summarized this way: the image cannot be understood independently from the technical mediations by means of which he who imagines, engages with the world. Images are part of a history, a technical history. In his words, likely to be found in Malafouris' lines on the material sign (Malafouris 2013; see next section), the mental image and the "object-image" (the concrete and technically made image) "cannot be separated from one another, no more than the signified and the signifying that defined, in the past, the two sides of the linguistic sign."

Let me explain that. Stiegler develops his theory of the imaginative schematism as a direct response to Kant's. Stiegler situates in the concrete and technical world the imaginative schematism that Kant located in the only transcendental and representational subject, on the side of pure interiority. Let us consider for a moment the conceptual and problematic context in which Kant first coined the concept of "imaginative schematism." In Kant, an object is represented and thought. A representation is made of two components: the sensuous intuition and the pure forms of appearance (space and time). The pure forms of appearance constitute the framework in which the unordered manifold of sensation is ordered, arranged. "Matter" ("that in the appearance that corresponds to sensation," A20/B34) is shaped, ordered in specific (spatial and temporal) relations. In this perspective, space and time, instead of being something that exists independently from the mind, out there in the world, are a priori conditions of sense-experience. They are contributions made by the faculty of sensation to cognition. Then, an object is thought thanks to the understanding, which Kant defines as the faculty of the mind that deals with concepts. The world of experience is formed in the application of the categories (or pure concepts) of understanding to the representation of sense-experience.

The question is to know how this subsumption of the intuition to the pure concepts of the understanding is possible, if the latter can never be discovered in the former. Indeed, according to Kant, the act of thinking lies in a kind of homogeneity or correspondence between the concept and the representation. For example, the round-shaped plate is represented and thought in my mind. The concept of roundness corresponds to what is represented in the plate. And the representation of the plate corresponds to what is thought in the concept of roundness. Only in that way am I able to think the round plate represented in my mind. Another example, "5," the material image of the number five, corresponds to the concept of this number. And vice versa, the concept of this number corresponds to material images like "5," "V," "five," ". . .," and so on. But the problem comes from the essential heterogeneity between sensuous representations and pure concepts. We do not see a concept directly in the sensuous representation. When I see the image "5," I do not directly think of the number five. I need to engage my understanding, my ability to deal with the concept five. In a few words, the mere representation does not exhibit the concept and the concept does not, in itself, contain the image.

To solve this problem, Kant resorts to imagination. He conceives of imagination as a mediating power, a faculty between sensibility and understanding. Imagination operates the correspondence between the manifold of intuition and the concepts of understanding. The imaginative and transcendental (transcendental as "condition of experience") schema is a procedural rule by which nonempirical concepts are associated with sense impressions. It is both a sensuous and an intellectual operation by means of which concepts apply to sensuous representations. Kant explains these operations in terms of a transposition of the pure concept of understanding in the form of time. The schema transposes the concept in the form of sensitivity, generating a sensitive image of the concept. In this case, the schema corresponding both to the concept and the image "five," consists in sequentially adding unity to unity in time. This kind of transposition in the sensitive form of time allows, for example, to think all the concepts related to the category of quantity: unity, multiplicity, and totality.

According to Stiegler, the schema opens up a new and problematic domain of sensitivity, as sensitive and particular as sensitive intuition, as intellectual and general as the pure concept. The question is, then, to know how to characterize the relation between the sensitive image and the schema. Kant says schemata are no sensitive images. But, if the schema transposes the concept in the form of time, if it does consist both in an imaginative procedure and in the sensitive image and transposition of a concept, what is it, if not a sensitive image? To put things in Stiegler's terms, "to what extent is a number like thousand possible, as a method consistent with 'a specific concept' . . . WITHOUT AN IMAGE?"

> The answer is clear: in NO way. The number always involves, in a way or another a capacity of tertiary retention (whether it is the fingers of a child, the body of a wizard, an abacus or an alpha-numeric writing system), which can alone enable us to count and objectivate.
>
> . . .

> This capacity has a history, in which in particular conceiving the number 1000 (thousand) became possible. (Stiegler 2001, p. 83)

By emphasizing the technical history of human intelligence, Stiegler radically breaks with Kant's transcendental perspective. If there is anything like an imaginative schematism allowing for the articulation between sensuous experience and conceptual thought, this schematism is technical instead of transcendental. Kant says schemata precede images, in the sense that there would be no meaningful image without the schematic and transcendental application of a concept. For his part, appropriating Simondon's concept, Stiegler argues that images and schemata co-emerge transductively. And they do so as a result of our technical engagement with the world. According to Stiegler, the making and use of technical devices is conditional with regard to the emergence of mind. Five fingers embody the number five. The making and use of an abacus affords a concrete, technical, and sensitive support for the conceiving process of numbers. Specific numbers are conceivable only on condition that they figure in numeration systems. These numeration systems are technical systems of symbolic manipulation, independently from which no number would be conceived and manipulated. In Stiegler's words, "there was a time, very new compared to the long history of mankind, where the number 1000 remained literally un-conceivable to the consciousness of man, who was not yet equipped to think it, where 1000 (thousand, or the image above, or 1111101000) was not yet elaborated" (Stiegler 2001, p. 87).

Image and schema are the two sides of a same reality. They constitute a historical process conditioned by what Stiegler calls the "epiphylogenetic structure," which refers, in Stiegler's terms, to the general system of "tertiary retentions." The term "retention" here, comes from Husserl. As a reminder, in Husserl, primary retentions concern what is happening here and now in consciousness. A primary retention is a mere presence to the perceptive and emotive flow. For example, listening to the BWV1011 prelude I mentioned in Chapter 4, I am present to the first sounding note, a doubled C. Secondary retentions are former primary retentions, populating consciousness under the form of memories. They pertain to imaginative memory and polarize the way we perceive, the very content of our present and future primary retentions. In this case, the double C gives me the key signature of the musical piece, in the light of which I am now listening and appreciating the next musical phrase. Stiegler adds to this equation "tertiary retentions." Tertiary retentions differ from Husserl's primary and secondary retentions by their technical and material nature. They refer to hypomnesic sedimentations, accumulated generation after generation in artifacts and allowing for the reactivation of past imaginative and conceptual experiences. They are memory supports, overdetermining primary and secondary retentions. They allow a transindividual and transgenerational process of psycho-social and technical individuation and sharing.

Let me explain this point more precisely. Inspired by Derrida's interpretation (1962) of Husserl's *Origin of geometry* (1936), Stiegler addresses the question of the originary and "proto-foundational" beginnings of conceptual domains. Husserl understands mathematical language and geometrical figuration as more than mere technical means of expression and consignation. Mathematical language and geometrical figuration express a meaning, says Husserl, to the only extent that they contribute to producing it. The mind does not exist as an absolute system, drawing mathematical and geometrical ideas from the inside. Instead, these ideas emerge in the core of technical practices. In a way that might seem paradoxical, science and ideas rise from consignation technics independently from which the very order of ideas and signification could not be achieved. No way to phrase this is better than Derrida's, whose analysis Stiegler explicitly extends:

> Writing is no longer only the worldly and mnemotechnical aid to a truth whose own being-sense would dispense with all writing-down. The possibility or necessity of being incarnated in a graphic sign is no longer simply extrinsic and factual in comparison with ideal Objectivity: it is the *sine qua non* condition of Objectivity's internal completion. As long as ideal Objectivity is not in a position to be party to an incarnation (which, in the purity of its sense, is more than a system of signals or an outer garment)—then ideal Objectivity is not fully constituted. Therefore, the act of writing is the highest possibility of all "constitution", a fact against which the transcendental depth of ideal Objectivity's historicity is measured. (Derrida 1962/1989, pp. 88–9).

In other words, the corresponding relation between sensitive images and concepts results from their technical co-emergence. The genesis of ideas lies directly in the making of technical systems of numeration or consignation, as well as in their concrete manipulation. In this sense, technics is conditional with regard to human intelligence. And it is "constitutive" of human intelligence in the sense that it allows both for the genesis of ideas and sciences, and for their transindividual and transgenerational transmission. Indeed, mathematical notation, or geometrical figuration, makes it possible to communicate mathematical and geometrical concepts, to share them through space and time.

Mathematical and geometrical notations, written signs in general, are sensibly experienceable. They act as external kinds of retentions, ultimately preserved or destroyed through time in that they exist in the form of artifacts. As such, it is always possible (when preserved) that they be intersubjectively experienceable.

In an Heideggerian vein, and prolonging Husserl and Derrida's analyses, Stiegler considers the individual being as always preceded by technics. Technics is "already-there," allowing individuals to inherit from a past they did not experience themselves. The central claim is simple. In Stiegler's words, the "who," that is, the subject with its specific mind, is always preceded by the "what," namely, the technical devices constituting a specific kind of subjectivity. In this sense, Stiegler proposes what we might call a technologization of imagination. Schematism is not a hidden art anymore, as is the case with Kant. Instead, it is observable in its constitutive technical traces.

This means two things. First, schematism is inseparable from the external and technical environment. Sensitive images and concepts co-emerge, in the sense that conceiving is at the same time producing a technical and sensitive image whose function is to serve as a tertiary retention in the process of psychic individuation. Tertiary retentions make possible the imaginative synthesis between past and present images. And they open up specific horizons of thought. The imaginative schema refers, then, to a concrete and technical practice independently from which no concept would exist. To put it another way, Kant's schema is a transcendental and operational condition of subjective experience. It is a rule by which the preexisting concept of the understanding is supposed to be transposed in the form of time, a form of sensitivity. From Stiegler's perspective, there is no transcendental operation. No transposition of any pre-existing and pure concept into any sort of sensitive image. For Stiegler, the genesis of images and concepts is concrete and technical and, above all, relational and transductive. Concepts do not pre-exist. Instead, they emerge through socially shared and technically shaped practices. The schema is not a transcendental rule transposing an already-given concept into the form of sensitivity. Rather, it is a technical process of image and concept co-genesis. Accordingly, sensitivity is prosthetic, technically extended into the external world. It is not internal anymore, neither is it limited to the organism's body and intrinsic properties. Instead, it is relational, in between internal and external processes, and mediated through technics.

Following Stiegler, the schema appears as the name of a question, rather than that of a solution. The question is to know how to explain the correspondence, the co-naturality between the sensitive image ("5," "V," "five" "IIIII," etc.) and its corresponding concept. Stiegler's answers consists in coming back to genesis. He explains the co-naturality in terms of technical co-genesis. Let me note that the general question here, if I may phrase it in enactive terms, is that of structural coupling. The question is indeed to know how to explain the ability a subject has to understand the meaning of her cultural environment. For it is not sufficient to say that technical objects enhance the living's capacities, or that they provide living beings with new imaginative, conceptual, and expressive horizons. As I emphasized in Chapter 3, the true difficulty is to account for the possibility of the dialectical and meaningful relation between the living being and its sociocultural environment. In the case of a mathematical notation for example, how are we to understand that the subject grasps the meaning of the material image "5," if it is not by means of a purely internal and transcendental imaginative process, isolated from the material and technical side of the image? Malafouris enactive theory of "enactive signification" will be extremely useful on this question. For his part, what Stiegler does in the wake of Simondon's works is, beyond merely postulating the dialectical relation between the subject and the world, explain it in terms of a technical ontogenesis of meaning. From the perspective of these two authors (Stiegler and Simondon) indeed, no doubt that if the object means anything for the subject, and if the subject accesses this meaning, it is in virtue of the co-origination and co-naturality of the subject and the object. No divide between the subject and the object. Both share a same origin, namely the individuating and technical relation between the individual and the milieu. They share a same nature, a same mode of existence, simultaneously technical and cognitive. And for both Simondon and Stiegler, the only way to think this dialectical relation between the subject and the object is to acknowledge its transductive nature. To put it simply, extending Simondon and Stiegler's perspectives in a language they do not use, we might say that cognition needs to be deeply reincorporated in technics and that we need to understand cognition as the technically shaped outcome of a technically mediatized relation with the world. And I tend to think that not questioning, like Simondon and Stiegler, the technical conditions of possibility of the dialectical relation between the mind and the material world makes it impossible for enactivists to get rid, once and for all, of representationalist and internalist accounts of cognition.

Second, to say that schematism is not a hidden art, amounts to articulating imagination to the history of its constitutive technological sources. This clearly resonates with Simondon's perspective again. Indeed, Simondon understands technology as a kind of reflexive psychology or archeology of the mind, a sort of a backward reflection on past imaginative and inventive processes through the study of given technological traces. As put by Jean-Yves Chateau quoting Simondon (1968/2005, p. 16), "'invention is the mental, psychological aspect of a specific mode of existence', that of the technical object (1968); it is the subjective correlate (*a parte subjecti*) of its concretization, of its concretizing genesis" (Simondon 1968/2005, p. 18). Put another way, invention, both a schematic and a material process of genesis, does not come from the subject, independently from the object. The object has its own mode of existence, its own concretizing genesis. In Simondon's terms, technical objects have their own "auto-correlation" (Simondon 1968/2005, p. 92). They exist as sets of operative modes and of functional relations between them. And imagination refers to a kind of participation whereby the genesis of the object equals the genesis of images and concepts. Technical objects do not enhance the living's capacities only. They also have their own intrinsic logic, according to which agents manipulate and inventively transform them. They have their own ontological status and deserve their own ontology (technology properly speaking; see Chapter 3).

For his part, Stiegler emphasizes the idea that technical objects enable, enlarge, and diversify the spectrum of human intelligence. According to him, any organological enlargement, that is, any change or multiplication of the available instrumental and technological devices, goes along with a change and a multiplication of the operative modes constitutive of human intelligence. In the case of musical creativity, for example, the numeric revolution which occurred in the 1960s marks an important turning point. Computational and digital technologies entered musical imaginative and creative practices. New technological ways for musical creativity emerged thanks to the digital and computational revolution. Through history, musical creativity constantly knew the same kind of organological changes. Each instrument knew progressive and historical transformations, both initiated by and initiating new musical, imaginative, and stylistic tendencies. Each instrumental and technological change enabled new ways to listen, to perform, and to compose. Interestingly, talking about "organological enlargement," Stiegler refers to a term proper to musicology: "organology." Organology is a branch of musicology. It refers to the study or science (*logos*) of the links between instruments (*organa*) and musical styles. Organology explains how new instrumental and technological devices give rise and reflect new musical, imaginative and stylistic experiences. It is an essential part of musical education, especially for those among musicians we call historic musicians.

Historic musicians specialize in early music, in medieval, Renaissance, or baroque musical styles, for example. They play on re-historicized instruments.

A baroque cellist, for example, uses bodily, fingering, or bowing techniques specific to the time period of the musical creation he interprets. The very essence of this so-called historical approach to musical interpretation lies in trying to grasp the stylistic principles governing the interpretation of past musical pieces, drawing directly from the study and practice of past or re-historicized musical instruments. The difficulty, for a contemporary cellist aiming at interpreting Bach's musical pieces in the early eighteenth-century German style—different from the late eighteenth-century Italian style specific to Boccherini's sonatas or concertos, for example—is that he is unable to go back in the past and listen to cellists of this time period and cultural area. The challenge is, then, to re-enact something approaching the interpretative style of this period, by (1) studying the writing technics of past composers, (2) playing on baroque musical instruments, and (3) applying instrumental technics typical of this time period. The baroque cello, with its own making and features, opens specific imaginative, expressive, and stylistic worlds. And the baroque cellist deals with these worlds, revealing and partly resurrecting them. For sure, this is not an easy challenge. But it is a meaningful one. And I aim, among other things, at enlightening the meaning of it, from the point of view of theoretical cognitive science. It is a way for me to pay tribute to my masters, Anner Bylsma, Wieland Kuijken, Alain Gervreau, Tormod Dalen, Emmanuel Balssa, Christophe Coin, Amandine Beyer, and Pierre Hantaï, among others. It is also a way to make cognitive science take into account what the art of such creative and imaginative genius says about imagination and technics.

As concerns Stiegler, I will conclude by emphasizing that any aesthetics is rooted in a functionalizing organological domain (Stiegler 2005, p. 199). In this case, musical images, with their aesthetical (both affective and formal) qualities, do not exist as pure reflections of material things. For sure, they live in the material life of sounding things. But not only. Neither do they originate in the mere affective, Merleau-Pontian body. Looking for aesthetical values in the body or in the material life of things is still missing something essential for the understanding of musical imagination. Namely, that musical imagination, beyond its materiality and embodiment, involves a kind of aesthetical experience, whose possibility lies in the existence of specific organological and cognitive domains. Composing and performing with a harpsichord differs from composing and performing with Ableton Live and a home studio. These cognitive and aesthetical experiences radically differ, and not only in their bodily affective and material dimensions. For sure, the materiality of the sounds, of the concrete and manipulative relation we engage in with those instrumental and technological devices, differs. But, above all, these technical and aesthetical experiences provide two different ways to bring forth the world, two ways of thinking, perceiving, and acting within it. A given aesthetics is not just a way to feel time, bodily, affectively, and materially speaking. It is a way to see the world, to engage with oneself and with the world, in unexpected ways, in ways the biological body of the organism does not provide by itself (see Chapter 7). It is a way to enact a specific world, by technically shaping and enabling specific kinds of enactive engagement. But what do the terms "enaction" and "cognitive" mean here? And is there a way to transpose Simondon, Ingold, and Stiegler's achievements in the language of enaction and cognitive science?

2. Imagination as a Technically Shaped Mode of Bringing Forth the World

Through technical engagement, things become meaningful. Technical objects enable specific cognitive modes of worldly engagement and acquire a meaning to the extent that they participate in its genesis. As Stiegler notices it, the challenge is to elucidate the nature of the processes that govern the co-genesis and articulation of neuronal and technical structures. For in this co-genesis and articulation between the neuronal and technical structures lies the very secret of constitutivity. In Stiegler's partly Simondonian terms, "there would be a double emergence of the cortex and the silex . . .. All the difficulty would be to reveal the complex (transductive) dynamics of this 'complex of Epimetheus.'" We all know the famous myth, as depicted by Plato in *Protagoras*. Epimetheus, lacking foresight, failed to give humankind the necessary attributes. By doing so, Epimetheus condemned human beings to find ways to make up for their physiological shortcomings. And invention became essential to human condition. In this perspective, technics appears as a prosthetic attribute, de jure limitless, by means of which humankind both repairs Epimetheus' fault and defines its essence. But how to characterize this essential articulation between neuronal and technical-material processes, from the point of view of cognitive science?

As I already emphasized, the pragmatist perspective I offered in Chapter 2 clearly resonates with the "integrative view" on cognition which proponents of 5E approaches (enactive, embodied, extended, embedded, and ecological), like Malafouris, call for (Malafouris 2007, 2013; Gallagher 2016, 2017, 2020; Di Paolo, Cuffari and De Jaegher 2018; Baggs and Chemero 2018; Van Dijk and Rietveld

2018). It echoes with many attempts during the last three decades to understand how technical devices, from tools we manipulate to social institutions (Gallagher et Crisafi 2009), shape cognitive acts like mathematical reasoning (Clark and Chalmers 1998; Kirsh 1999), calculation (Wilson 2004), perception (Lenay 2006), perceptive guidance (Hutchins 1995b), and memorization (Donald 1991).

And it provides perspectives on imagination that echo recent works on imagination in the field of 5E approaches (Van Rooij, Bongers, and Haselager 2002; Malafouris 2007; Ingold 2013, 2014; Hutto 2007, 2015; Rucińska 2014, 2016; Gallagher 2017; Hutto and Myin 2017; Van Dijk and Rietveld 2020). Elaborating on Lambros Malafouris' Material Engagement Theory (2013), (Hutto 2015) defines "kinematic imagination" in terms of prelinguistic instrumental thinking. Instrumental thinking involves an imaginative and nonverbal capacity to mentally rehearse memories of action-perception patterns. Early humans of the Middle Paleolithic, those capable of instrument making (see the example of the Levallois flake), were able to engage in imaginative rehearsals consisting of "visual-motoric perceptual reenactments" (Hutto, 2007, p. 84). As Medina (2013, p. 229) emphasizes, this enactive imagination has to do with our constant embodied and practical engagement with things. It consists in the imaginative reenactment of "our experience as engaged actors acculturated into social practices" (Medina, 2013, p. 319).

As I emphasized, Gallagher outlined the main features of an enactive-ecological approach to imagination (Gallagher 2017). Drawing upon Ryle's idea, according to which imagining is a doing (Ryle 1949), Gallagher defines imagination in embodied, enactive, and ecological terms, as a form of active engagement with affordances (Gibson 1979), that is, with action possibilities (Gallagher 2017, p. 193). When imagining how a tune goes, for example, we "make ready for those notes in a hypothetical manner" (Ryle 1949/2009 p. 245). Not that we manipulate any mental representation of a tune going this or this way. Rather, we actively engage in the simulation of a given possible action, that of humming. "We do what we would do if we were going to hum the tune, but simply stop short of actual humming" (Gallagher 2017, p. 193).

In a very pragmatist verve, Gallagher notes that, defined this way, "imagination is not something that happens first in the head; it's rather something that involves embodied action, using toys, props, artifacts, instruments, and so on" (Gallagher 2017, p. 193; see also Rucińska 2014, 2016, on playacting; see Chapter 6). Gallagher even assumes that the extended nature of imagination understood with pragmatism, in terms of bodily and practical engagement with affordances exhibited by material things, "needs to be the starting point for the analysis of imagination" (Gallagher 2017 p. 193fn). In this sense, Gallagher takes Malafouris' Material Engagement Theory to be nothing less than the "starting point for an enactive study of imagination":

> Hutto makes this clear when he links his radical enactivism with material engagement theory (MET) and the work of Lambros Malafouris (2013). I think this needs to be the starting point for the analysis of imagination. Engagement in pretend play, or in working with material things, such as stone tool making, is where the imagination starts. "Stone tools are not an accomplishment of the hominin brain, they are an opportunity for the hominin brain — that is an opportunity for active material engagement." (Malafouris 2013, p. 169; Gallagher 2017, p. 193)

Like perception (Chemero 2009), imagination is embodied and situated. And it is temporally extended (Van Dijk and Rietveld 2020) in the sense that these action possibilities, constitutive of our imaginings, are enacted through the history of our interactions with materials, as well as with other people.

In the next section, I will take Gallagher's recommendation seriously and bring some water to the mill of this enactive, embodied, and situated account of imagination. This will give me the opportunity to go beyond the limitations of what I said in Chapter 4 about the material and technical nature of images. To put it very simply, the stake is to offer a theory of the articulation between cognition and the sociocultural environment, and to draw conclusions about the nature of cognition in general and of imagination in particular. As I emphasized, the enactive approach recently made it a priority to think this articulation. Beyond the limitations of its overly body-centered formulation, the enactive approach needs to take the environment more seriously. We need to understand how minds, and the material and socially shared world, are enmeshed and related. We need to explain through what mechanisms those linkages are made effective.

2.1. The Technical and Material Constitutivity of Cognition

Malafouris emphasizes from the start the necessity not to adopt an instrumentalist view on technics. Instead of an a posteriori means for the material consignation or expression of pregiven mathematical ideas, technical devices are material mediations by means of which, only, mathematical capacities, concepts, and ideas can emerge. In Malafouris terms, Mycenaean cognition "lies out there in the world," for the reason that "it was enacted *through*, rather than written *upon*, the Mycenaean tablets" (2013, p. 68).

But how to explain that? If Mycenaean tablets were more than mere consignation devices, what role do they play concretely in both the emergence and the existence of the Mycenaean mind? More fundamentally, to what extent are these Mycenaean tablets more than external means for purely internal and representational processes? After all, why not endorse a causal coupling explanation, in the manner of Andy Clark and David Chalmers (Clark and Chalmers 1998)? According to them, mere things actively participate in our cognitive life. Technical devices are cognitive, in the sense that they are constant and constitutive *functional* parts of cognitive processes. Famous Otto's notebook realizes the same cognitive function as Inga's neuronal and internal memory. Religious or technical artifacts enable us to use biological skills in ways that would be impossible otherwise (Clark 2010). Here, the ontological difference between internal and neuronal processes on the one hand and external and material-technical ones on the other does in no way invalidate the idea that technics is constitutive of cognition—from a functional point of view. But, at the same time, to say that technics is constitutive of cognition in Clark and Chalmer's perspective in no way invalidates the representational theory of mind:

> We should not be too quick to reject the more traditional explanatory apparatuses of computation and representation. Minds may be essentially embodied and embedded and *still* depend crucially on brains which compute and represent. (Clark 1997)

Then, does Malafouris provide a truly alternative way to think the technical constitutivity? Is there a way for us to think the technical constitutivity of imagination, from the point of view of cognitive science and in a way irreducible to Clark and Chalmers' representationalist externalism?

Malafouris' thesis can be summarized in a very simple formula, whose explanation, however, takes more than a simple phrase: cognition lies in the material structures and settings of *a practical, both manipulative and productive activity*. Malafouris accounts for the genesis of meaning, of concepts, ideas, and cognitive capacities, in terms of structural coupling, that is, of a genetic and essential articulation between cognition and technical-material structures. He does so by developing a stimulating and nonrepresentationalist theory of "enactive signification." The stake is to understand how material structures become meaningful by themselves, independently from the denotative function of linguistic signs, *as well as from any representational process*. In its Saussurian and classical definition, a linguistic sign consists in a conventional relation between a signifier and a signified. It is a two-sided psychological relation between a vocal or visual element (the signifier) and a concept (the signified). For example, the signifier "5" arbitrarily corresponds to its concept. Different conventional notations like "five," "fünf," "V," and so on correspond to the same concept.

Malafouris's claim is that "the world brings meaning by itself, in ways that language cannot" (2013, p. 95). Moreover, the linguistic sign must be understood as nothing but the last result of an enactive and technical, transhistorical and transindividual, and, finally, nonrepresentational genesis of meaning. The emergence of the intrinsic meaning of the linguistic sign lies in a kind of material engagement whereby enactive, prelinguistic, and nonrepresentational kinds of signification emerge.

As Malafouris emphasizes, what we are lacking is a precise explanation of the cognitive mechanisms able to account for the material constitutivity of enactive signification. For his part, he provides this explanation in terms of a process of embodied "conceptual integration" responsible for the co-substantial symbiosis and simultaneous emergence of material structures, significations, and cognitive capacities. As concerns the cognitive ability of counting for example, Malafouris draws from Peter Damerow's work on the use of clay tokens in the preliterate period (1988, 1998). He argues that material structures like clay tokens, impressed tablets, or pictographic tablets, with their intrinsic physical qualities, mediated the very emergence of mathematical concepts. Clay tokens and impressed tablets must be understood as epistemic artifacts enabling counting. Counting is not a faculty magically offered by the naked brain. Rather, it is the outcome of a material and technical engagement with the world. It is enacted through material engagement.

Clearly, we recognize here the Stieglerian idea that technical devices enable us cognitively speaking. But we still do not know how such a theory of enactive signification overcomes the supposed limitations of a causal coupling explanation. Following Damerow, Malafouris says, "the initial emergence of the concept of conservation of quantity is tied to *the substantive reality and concrete use of clay tokens* and not to any pre-existing cognitive skills of an arithmetical nature" (Malafouris 2013, p. 111, emphasis added). Regarding these "substantive" bits of clay, Stiegler would talk about tertiary retentions, overdetermining and making possible new kinds of retentional syntheses. Clark and Chalmers would talk about external supports for the development of non-pre-existing cognitive domains.

For his part, instead of using Kant and Husserl's languages, and explicitly rejecting the language of classical representationalism, Malafouris adopts a pragmatist, nonrepresentationalist, and enactive language. According to him, indeed, "cultural knowledge and innovation are not intracranial processes; they are, rather, infused and diffused into *settings of practical activity*" (2013, p. 116, emphasis added). The fundamental idea is that, instead of being the external means of internal and representational processes, clay tokens or tablets constitute new numerical "habitus," that is, new cognitive operations. But these cognitive operations are not to be understood in terms of internal and representational processes. Instead, and this is Malafouris' original claim, they are practical, manipulative, material in themselves! They emerge in specific technical contexts, depending on how material structures and settings shape our embodied engagement with the world. There is something like a direct physical grasping of numbers. Numbers, as concepts, *emerge through and directly consist in* the physical manipulations of technical-material devices.

Let me be more specific about this crucial aspect of the argument. Malafouris grounds his theory on Edwin Hutchins' notion of "material anchoring" (2013, p. 104). Material anchoring refers to the cognitive projection from a conceptual domain to a material domain. In the case of counting, basic numerosity is objectified through the materiality of the clay tokens. A familiar domain of meaningful experience is projected into a material, a priori meaningless material structure. The vague concept of unity, as it is experienced in the embodied and socially shared use of one jar of oil, or of one basket of grain for example, is projected in the domain of clay, by means of a one-to-one correspondence. The vague concept of unity is projected in the clay token, and becomes physically objectified, manipulated, and combined to other objectified and manipulated unities. The cognitive ability of counting emerges, then, as the result of material anchoring. The material structures make counting, originally a vague and approximative conceptual process, a concrete and practical one. Counting becomes an embodied and practical process of technical manipulation. The manipulation of numbers in the form of physical objects, clay tokens, tablets, and so on enhance and tighten the conceptual capacity, making counting a less approximative, a more complex cognitive process. No way to say it better than Malafouris himself:

> What essentially happens in those cases, put in very simple terms, is that the vague structure of a flexible and inherently meaningless conceptual process (e.g., counting), by being integrated via projection with some stable material structure or thing, is transformed into a perceptual or physical process. However, perceptual operations embody a spatial logic and thus can be directly

> manipulated and explored in real time and space. Thus, the process becomes meaningful. (2013, p. 105)

Put another way, the manipulability of clay tokens, as bits of matter situated in a perceptible physical space, transforms a vague concept of unity into a meaningful and manipulable one. The cognitive domain of practical and social engagement with commonplace objects, like those we use in our everyday life, merge with the domain of spatial manipulation. We clearly see that it is not question for Malafouris to provide a classical computationalist explanation of counting. No causal coupling between internal, representational, and computational processes on the one hand and material, technical, and external processes on the other. And there is more to it than a mere functional equivalence between neuronal and technical-material processes. Again, counting, a cognitive process, appears in Malafouris as a concrete, embodied activity of spatial manipulation. It is engaging with technically shaped and shaping material structures, independently from which no mathematical concept or ability would be possible. Put another way, counting is bodily engaging with the technical device, a kind of engagement whereby a specific conceptual domain emerges.

Let me insist. To say that cognition is a practical activity does not amount to saying that it is a representational one, made possible by the practical manipulation of technical devices. Instead, cognition appears as a concrete manipulation in itself, by means of which meaningful domains of experience are objectified. Material anchoring appears, in this perspective, as a process of objectification and rationalization of embodied and affective experiences, through a kind of technically constituted discipline of the body. This process, in Kant's words, realizes the ontological merging of sensitive experience and the concept. Worth noting, it is going a step further than the functional identity thesis. It is accounting for this functional identity in the terms of an ontological blending, of an "anchoring blend," as says Malafouris, between two nonrepresentational domains of embodied experience. In this sense, Malafouris explicitly refers and endorses Clark and Chalmers' extended mind thesis (Malafouris 2013, p. 104). He seems to agree with the functional identity thesis. But he explains it in nonrepresentationalist terms, and of (1) a technical ontogenesis of cognition and (2) an ontological identity between cognition and material-technical engagement.

In the wake of it, Malafouris offers an enactive working hypothesis to answer Stiegler's question about the co-emergence of the cortex and the silex. How to think their transductive relation? Following Malafouris, the intracranial processes involved in this kind of material, meaningful, and enactive processes must be understood in terms of, let me say it this way, "biological prostheses" of the material living world. Malafouris does not use this formula. But I guess it properly expresses his idea, by difference with Clark and Chalmers' intuitive and apparently identical claim that brain structures are affected by the use of technical devices. "Individual learning, Clark and Chalmer say, may have moulded the brain in ways that rely on cognitive extensions that surrounded us as we learned" (Clark and Chalmers 1998, p. 12). For his part, drawing from the works of Kelly and Garavan (2005) and Poldrack (2000), Malafouris takes the process of engaging with material structures, of enacting through them meanings, concepts, and cognitive abilities, to cause an "extended neuronal reorganization" (2013, p. 115). At first sight, there is no difference. However, in terms close to Malafouris heart, it is a question of recasting the boundaries of mind beyond the skull, right in the manipulative hand, in between internal and material processes. It is a question of defining the "mental" in terms of this relational process, in between the internal and the external, in between the neuronal and the technical. This operates a kind of conceptual decentration, from the side of internal mind-brain and supposedly representational processes to the side of nonsubstantial, relational, and material processes. The mind is not, then, in the brain. It lies "out there in the world," a living material world of which we are nothing but a constitutive and living part. In this sense, we might accept Clark and Chalmers' stance that external devices reorganize the brain, without postulating the existence of internal representational processes.

Let me summarize the argument. It is not question to discuss Clark and Chalmers' claim that technical devices shape the mind, reorganize the organism's brain, and enable us cognitively speaking. There is no question of discussing Clark and Chalmers' extended mind thesis. Rather, Malafouris aims at accounting for this extension of mind in terms able to explain the radically embodied, material, and technical nature of mind. Beyond the mere technical ontogenesis of human intelligence, the mind is in itself a pluri-modal, bodily, material and technical engagement with material structures. The mind is not something like a substantial and world separated pole of thought, perception, and action. It is, rather, a practical activity in itself, an ongoing process of perception and action. Technical devices, in this sense, are no static means for representational processes. Instead, they are living constituents of enactive and non-representational signification processes.

In this sense, saying that intracranial processes are biological prostheses of external and material processes amounts to reversing the classical view according to which technical objects are prostheses of biological and brain processes. It amounts to emphasizing the other side of the dialectics, the essential and, above all, constant relativity of the organism and of the brain to the external, socio-material world.

Finally, it is worth taking seriously Malafouris's enactive formula, according to which material engagement brings forth the world. What is a world? It is not just the external environment. We often speak about someone's world. My world is not that of a North Korean, and so on. Depending on material, economic, social, technological, political, religious, and so on factors, worlds multiply in different and often opposite directions. To speak naïvely, the world is a "subjective" thing. But "subjectivity," here, might be something less simplistic than a set of internal and "mental" representations (where "mental" means internal and computational). Following Malafouris developments, "world" means (1) the concrete, material (neuronal and technical), and meaningful relation whereby the cognitive agent and the cultural world transductively co-emerge—in this sense, it is an ongoing enactive signification process; and (2) a set of cognitive procedures that transform themselves through time, along with the history of material culture. In particular, these cognitive procedures transform along with the history of the material-technical structures that couple the living being and the environment. By "cognitive procedures," here, I mean the ways we perceive, act, and think. That is, the practical routines we acquire through technical and material engagement—procedures by which the enactive signification process occurs. In this sense, being a cognitive subject, or a mind, amounts to being a prosthetic body, engaging with things through cognitive projection or material anchoring. It is being a world, in the sense that the mind lies out there as a material world in formation.

2.2. The Technical Constitution of Enactive Imagination

Malafouris thematizes the question of imagination in recent stimulating texts (Malafouris 2018, 2019a, 2019b, 2020a, 2021a, 2021b, 2021c; see also Koukouti and Malafouris 2020a, 2020b). From the start, his understanding of Paleolithic images, those early pictures we find on Chauvet Cave walls, brings interesting perspectives on imagination. A classical internalist and representationalist conception of perception would say that, when perceiving those images, we mentally build a representation of them. What our experience of seeing really consists of, according to this classical and widely shared conception of perception, is the mental representation of the retinal reflection of the material image we see.

We perceive, as Malafouris puts it, the representation of the representation of a representation. But according to Malafouris, those material images, painted on the walls, act as material structures whereby we engage with things in new and unexpected ways. Radically anti-representationalist and externalist, Malafouris understands those pictures or material images as specific modes of probing the world. More specifically, those pictures afford new ways to explore, perceive, act, and think. And those ways are made possible through the concrete practical activity of painting:

> I propose that images like the ones we see, already 30,000 years before present, at the caves of Chauvet and Lascaux *before* and *beyond* representing the world they first *bring forth* a new process of acting within this world and at the same time of thinking about it. This thinking however, should not be understood—at least not in the first instance—as that of the "higher level" abstract or symbolic type. This thinking should be understood in the more basic "lower level" sense, namely, as a new form of active sensorimotor engagement (O'Regan 1992; O'Regan and Noë 2001; Hurley 1998). It should be understood as a new form of perceptual learning on a par with the 'bringing forth' or "bringing out" of a figure by embellishing the natural formation of the rock. Or, alternatively, a practice-induced change in the human ability to perform certain 'unnatural' perceptual tasks. (Malafouris 2007, p. 295)

Once we acknowledge, with O'Regan and Noë, that perception is not a process of representing but of a process of probing the outside world, the painted image appears, in Malafouris terms, as a "perceptual device." Whether we speak of painted images, or of instrumentally performed and sounding images, or of visual and moving images like those a nineteenth-century optical theater (see Hayden White 1973) or a modern camera makes possible—in all cases, material images are prosthetic extensions and transformations of perception, action and thought. They enable, enhance, diversify the way we perceive, the way we enact the world. They create specific dispositions and sensibilities:

> Cultural knowledge and innovation are not intracranial processes; they are, rather, infused and diffused into settings of practical activity and thus they are constituted by experience within these settings through the development of specific sensibilities and dispositions, leading people to orient and think about themselves within their environment in specific and often unexpected ways. (Malafouris 2013, p. 116)

Material images change our minds. To use Malafouris words, oil painting, playing the cello, and manipulating a camera refer to different kinds of "skillful interactive engagement" with the world (*Ibid.*). As a cellist or a moviemaker, I am disposed to engage with specific kinds of image perception and production. Malafouris applies this to the case of Paleolithic images. In a very Stieglerian way, he argues that "the question to ask about paleolithic imagery is not 'what kind of mind was needed to make those images?' but instead 'what kinds of minds are constructed by perceiving—and I would add "by producing"—those images?'" (Malafouris 2007, p. 295). As Stiegler would say, the "who," with its specific mind, is always preceded by the "what." Painting, playing a musical instrument, and so on, as ways to produce material images, consist in the technical ontogenesis of specific minds.

I will thematize this definition of imagination in terms of technically enacted dispositions and sensitivities in Chapter 7, when specifying how I positively explain imagination. For now, let me say that, with Malafouris and Stiegler as well, we reach the conclusion I have sketched out at the beginning of this book in the terms of Leroi Gourhan. Again, let us not make any confusion between "effect" and "corollary." The world is not an effect of the brain, and the brain is not an effect of the world. Rather, both co-individuate in the form of indivisible cognitive activities that are intrinsically biological and technical at the same time. The brain and the world are extensions of each other. The brain is the physiological extension of external processes independently from which no cognitive or imaginative act would be accountable. In this sense, to say that the brain is an internal process makes no sense. The brain is "as much a cultural artifact as a biological entity" (Mithen and Parsons 2008, quoted in Malafouris 2013 p. 45).

Worth noting, the notion of imaginative schematism literally grounds Malafouris's theory of constitutivity, in the sense that it is key to the understanding of cognitive projection and material anchoring. Cognitive projection refers to a process of integrative conceptual mapping, by means of which proponents of the embodied mind thesis understand cognitive enablement. A cognitive cross-domain mapping enables us to enter new and emergent cognitive domains. A source domain is "projected" into a target domain. For example, we usually say that "time goes by so fast!," that "Christmas is gone," and so on. The source domain of space and the target domain of time are projected into a blend. Time is thought and understood in terms of motion in space. But where did time go? From where, to were, and going through what locations? Usually we take these metaphorical expressions to be mere figures of speech. But from the perspective of embodied cognition, "metaphoric projections" are constitutive of thought. They enable us cognitively speaking and structure our conceptual understanding. In this case, the source domain of our affective and vague experience of time becomes objectified through our concrete and embodied experience of space. A third conceptual domain, that of objectified time, emerges from the projection, and is irreducible to each of its constitutive dimensions.

Those metaphoric projections occur by means of "image-schematic structures" (see, for example, Fauconnier and Turner 1998; Lakoff and Johnson 1999). Image-schematic structures refer to "recurring patterns in our sensorimotor experiences and perceptual interactions, such as source-path-goal, center-periphery, experience of bounded interiors, the gravity vector, balance and equilibrium, and force dynamics" (Malafouris 2013, p. 61). As an example, the container schema comprises a set of topological relations between an interior, a boundary, and an exterior. A metaphoric projection of this image-schematic structure gives rise to new and emerging conceptual spaces, like those of categories understood as containers of ideas, and bodies as containers of souls (Malafouris 2013, p. 102).

According to Lakoff and Johnson's embodied perspective, "our brains are structured so as to project activation patterns from sensorimotor areas to higher cortical areas." Conceptual knowledge is embodied, mapped within our sensory-motor system. In other words, the imaginative schemata through which concepts acquire their sensitive form—in fact, the very essence of concepts—refer to such patterns specific to sensorimotor and embodied experience. From a neuronal point of view, in Gallese and Lakoff's words (2005), concepts are neuronal, multimodal, and "functional clusters." A cluster is a "cortical network that functions as a unit with respect to relevant neural computation" (2005, p. 6). They integrate sensory areas like those involved in vision, touch, and hearing, with motor control and planning areas. And they provide the very structure and non-propositional "content" of concepts. An "embodied neural theory of concepts" (Gallese and Lakoff 2005, p. 12), consists in emphasizing the embodied and sensory-motor dimension of the neural structures that support conceptual thought and neural computation.

Presented this way, these imaginative schematic processes are internal. They occur in people's heads, under the form of neuronal and phenomenal experiences. They are "mental" in the traditional and internalist sense. And imaginative schematism lies in the body, at least in the neuronal patterns corresponding to our embodied experience. Hutchins grounds his theory of material anchoring in Fauconnier and Turner's works. For sure, he goes a big step further into externalism, in the sense that he takes conceptual models to crystallize in material structures. But he remains loyal to a representationalist language which Malafouris interestingly abandons.

According to Hutchins indeed, material structures are immutable. This enables the stabilization of the conceptual relations we represent in our mental space. Blended with external material structures, the mental and representational space stabilizes (Hutchins 2005, p. 1562). For his part, Malafouris does not deny the implication of intracranial and neuronal processes. But he defines schematism as being "infused and diffused into *settings of practical activity*" (2013, p. 116, emphasis added). Accordingly, the whole body (not only the neuronal structures involved in embodied experience) and material structures and the way they behave (the cello and its concrete material and sounding behavior, for example) are ontic parts of the schematic process. Again, it is not about denying the existence and constitutive role of neuronal and phenomenal (nonrepresentational) processes. It is just emphasizing that schemata occur out there in the world and, above all, in the concrete relations we engage with material things. This way, Malafouris breaks away from Lakoff and Johnson's embodied cognitivism as well as with Hutchins computational externalism. Internal and neuronal processes, which do not consist in representational processes, do not exhibit the whole ontological richness of imaginative schematism. Instead of being a mere internal and representational process, schematism is practical and is material in the wide (especially in the technical) sense of the term. This, as Koukouti and Malafouris emphasize (2020a, p. 39), leads to (1) rejecting any rigid distinction between the real and the imaginary and (2) changing the temporal and spatial boundaries of imagination—two theoretical choices I shall thematize in the following and final part.

Conclusion

Stiegler and Malafouris develop converging philosophical perspectives. Both resonate well with Simondon and Ingold's philosophies of imagination and creativity. Situating himself in Simondon's ontology of transductive relations, Stiegler explains imaginative schematism in the terms of a technical originarity and constitution of imaginative schematism. Against Kant's transcendentalism and rejecting classical cognitivism and neurocentrism, Stiegler takes technical object to open up conceptual and imaginative domains. In a language which Stiegler does not practice, that of cognitive science, Malafouris emphasizes along the same path, that such a schematism can be explained in noninternalist and nonrepresentationalist terms. Schematism needs to be thought in terms of a constant process of worldly engagement, where technical objects and devices enable us, cognitively speaking.

So be it. But now comes the obvious objection: Am I truly talking about imagination here? Clearly, I am not dealing with what we usually call "mental images." Until now, I have talked about material images and their technical production. Those technically made material images, with their production, bring forth new modes of enactive engagement. They create a disposition to perceive and produce material images, through the mediation of technical devices. Fine. But if I close my eyes and stop any technical engagement right now, I can freely visualize fanciful and nonexisting things. Ontologically, in the very moment I possess and manipulate them, they are neither technical nor external in themselves. They correspond to neuronal processes and consist in private phenomenal experiences.

Does this entail that there is a domain of representational images, irreducible to the kind of nonrepresentationalist perspective I have developed until now? The question rises and resists.

Chapter 6 will be the place to address and answer it and to show the true potentialities of Malafouris' enactive and externalist approach to imagination in terms of "imaginative praxis" (Koukouti and Malafouris 2020a, p. 43). More specifically, I will show that the question is not to know if taking imagination as a mode of concrete engagement, as I did, is a wrong way to proceed because it prevents us explaining abstract, offline imaginative experiences. Instead, the question is to be envisaged the other way round. Indeed, what if abstract cases of imagination did not exist? What if talking about abstract or offline cases of imagination and, more generally, of cognition were the symptom of a representationalist and internalist ontological choice that has no good epistemological reasons to be made, the result of the mereological fallacy I denounced in Chapter 2o? What if we were able to explain private imaginative experiences that take place in the present, which we usually consider as offline of abstract, within the framework of an ecological ontology that tells us that any cognitive, imaginative, and creative act is online?

Part Three

6

Situating Imagination

1. Approaching Experiential Imagination Differently

As I have shown in Chapter 1, the current literature characterizes experiential imagination as an internal and representational process. According to the accepted representationalist view on experiential imagination, imagining playing Beethoven's Waldstein sonata on the piano involves three types of imaginings: (1) "imagining seeing movements of one's fingers on the keyboard"—which refers to sensory imagination; (2) "imagining having a proprioceptive experience of these movements"; (3) "imagining playing the sonata." Imagining playing an instrument involves recreating the perception of bodily movements, in the form of mental sensory and proprioceptive images or representations (Dokic and Arcangeli 2015, p. 4). It also involves recreating an agentive experience. Motor imagery, Goldman claims, is "the representation or imagination of executing bodily movements," and has its counterparts "events of motor production, events occurring in the motor cortex that direct behavior" (Goldman 2006a, pp. 157, 158, quoted in Dokic and Arcangeli 2015, p. 4).

It is assumed that proprioceptive and motor imageries enable us to access the subjectivity of imagined experiences. Subjectivity, say Dokic and Arcangeli (2015), refers to interiority, and objectivity to exteriority. Sensory imagination is "objective" in the sense that it takes roots in the perception of the external world. Experiential imagination, by contrast, is subjective in that it is rooted in interiority, that is, in feelings, proprioceptions, and our sense of agency. As I emphasized in previous chapters, such classical oppositions between interiority and exteriority and between subjectivity and objectivity have no place in the pragmatist and relationalist ontology I have developed until now.

I will come back to this question of subjectivity in Section 2.1. For now, let's just consider this widely shared idea that, thanks to proprioceptive and motor images, one imagines playing the Waldstein sonata as "her" experience, an experience she feels in her own flesh. Following this line of thought, one could say, for example, that a singer does not learn how to sing a melody drawing only from visual images of her teacher singing the same melody or from auditory images of her teacher's voice. Rather, she associates to such visual and auditory memories the representation of her own bodily and vocal actions. She builds her own bodily and vocal expression in the form of experiential imaginative rehearsals.

At first glance, two questions arise. First, to what extent is experiential imagination embodied? It is said that experiential imagination enables us to engage in the subjectivity of imagined experiences, to feel imaginative experiences "in our flesh." Musicians and music lovers, according to their own capacities, understand music by means of mimetic motor imagery and action (Cox 2016; see Chapter 1). But how do motor imagery and motor action relate to each other? Do we need to say that motor imagery is a mental representation or image of motor action, independent from concrete action and from bodily engagement? By extension, and this is a second question, are neuronal processes, taken independently from bodily experience, enough to explain the meaningful character of such imaginative experiences? Do neuronal processes have any representational power independently from bodily experience?

No doubt that, when imagining playing Beethoven's Waldstein Sonata, I picture myself playing the piano. No doubt that imagination, in this weak and general sense, is representational. As I emphasized with Goodman in Chapter 2, no one would deny that we have a private imaginative life, that we are able to imagine, to represent, in the intimacy of our closed eyes, fictional situations, nonexistent things, and so on. But why should we think of these imaginative experiences and capacities in the terms of internalism and representationalism? Why should such cognitive acts like imagining or representing (as well as conceiving, anticipating, and so on) be made possible and realized by mental representations, that is, content-carrying neuronal structures, physiological processes enabling us to imagine and to represent fictional situations, actions, and things, enabling us to access their intrinsic meaning and to relate to oneself and to the external world in an intensional manner? Why should we assume the existence of such powerful content-carrying neurophysiological processes, the so-called "mental representations," when representational and imaginative acts and capacities could be explained, as I suggested in Part Two and in the wake of the pragmatist perspective I developed in Chapter 2, in terms of bodily processes and individual-world relations? Do mental representations help us explain something about imagination that we could not otherwise explain? Or, to the contrary, do representationalism and internalism prevent us to think of experiential imagination more efficiently?

To complete the epistemological and ontological criticism of internalism and representationalism I developed in Chapter 2 with pragmatism and Wittgenstein, let me make a little detour through Hutto and Myin's enactivist criticism. According to them, asserting that neurophysiological processes function as mental representations and are intrinsically contentful, amounts to assuming something impossible to prove and to understand, that meaning and content can be produced, possessed, and manipulated on the basis of purely natural resources. The question, here, is double.

1. First, it is the question of the origin of meaning. How to understand that things are meaningful for organisms? This is the question I addressed and answered in Chapter 3 in Simondonian and enactive terms.
2. The second question, known as the "Hard Problem of Content," is to know "how creatures that begin life with only basic contentless minds could ever come to have minds of a content-involving sort" (Hutto and Myin 2017, p. 123). It is to know how to naturalize content. And for Hutto and Myin, it is simply impossible to explain in a naturalistic framework, on the basis of purely natural relations of information or covariation for example, the emergence of content.

Content, at its simplest, refers to states of mind that possess correctness conditions. To be in such contentful states of mind is, then, "to take ('represent', 'claim', 'say', 'assert') things to be a certain way such that they might not to be so" (Hutto and Myin 2017, p. 10). Central to mainstream cognitive science and to many information-processing accounts of cognition, indeed, is the idea that mental representations have semantic content and can be evaluated with respect to properties like truth, accuracy, consistency, or appropriateness (Fodor 1981; Pylyshyn 1984; Von Eckardt 1993). These theories explain cognition in terms of computation, where computation means ruled-governed sequences of semantically evaluable objects (see Pitt 2020). For example, the belief "it is snowing" can be true or false, depending on whether it is snowing or not. The assertion "snow comes when it's hot outside" is inconsistent, and the desire to "eat the Eiffel tower" is inappropriate.

In Hutto and Myin's perspective, basic minds are contentless. Basic minds, or basic cognition, here, refer to all cognitive activities that do not involve the use of language and of cultural symbol systems. Catching a ball, situating oneself in a space, or keeping track of another's gaze, for example, appeals to dynamic and situated embodied interactions and engagements with the world. Basic cognition is intentional in the sense that it is world-directed, that it meaningfully relates to the world. But the intentional relation here, whereby the organism meaningfully engages with its environment, is nonrepresentational, does not involve the manipulation of content. It does not involve relating to things as being thus or so, as being true or false, accurate or nonaccurate. Rather, it consists in patterns of organism-environment interaction. To be a meaningful object for, let's say, a frog, is not to be a representation of a thing, stored in that frog's representation box. The object is the outcome of a contentless process of objectivation, that is, a process through which the organism targets and constitutes an object of lived, that is, affective and action-oriented, experience:

> How should we understand the way of being related to a worldly offering such that it is an object for the organism or system and not a mere thing? The times philosophical example—the case of the humble frog and his crude feeding habits—provides a convenient way of explicating REC's nonrepresentational take on objectivation. Frogs are inclined to lash out with their tongues when presented with small, dark stimuli that move in ways that are sufficiently like the movements of flies, their traditional prey. Thus frogs reliably respond to a range of different things that exhibit this signature behavior—and that list includes many things that are not flies, such as BBs or shadows moving in the right way. (Hutto and Myin, 2017, p. 102)

In Hutto and Myin's perspective, then (see also Hutto and Satne 2015), in a way very close to Merleau-Ponty's Goldsteinian and phenomenological conception of intentionality (see Chapter 3), intentionality is pre-reflexive. Certain states of organisms have the biological function of targeting objects and situations without contentfully representing them.

By contrast, then, how to explain the emergence of content? How to explain that organisms can, as it is the case for "integrated" humans (see Chapter 3), relate to things in a contentful way? According to Hutto and Myin, contentful states require social cognition and public symbol systems. Hutto and Myin make a distinction between basic (contentless) and high-level (contentful) cognition. Without entering in the detail of this distinction here, a distinction I intend to discard in next section, basic social cognition, in Hutto and Myin's perspective, refers to situations in which individuals target each other. Such basic social cognition is basic in the sense that individuals target each other without representing each other under descriptions with correctness conditions. In order to become content-involving, social cognition needs something more, namely, "the development, maintenance and stabilization of practices involving the use of public symbol systems through which the biologically inherited cognitive capacities can be scaffolded in particular ways" (Hutto and Myin 2017, p. 145). Claim-making practices for example, whereby individuals respond to things by representing them as being thus or so, are cultural practices. Such practices take place in culturally constituted contexts, according to norms of correctness that are socially constituted, through the use of symbol manipulation systems.

The idea is double. First, it is to explain meaning in enactive terms, in terms of embodied and relational individual-world dynamisms. Second, it is to relieve natural processes from the responsibility of content. Natural processes do not explain, by themselves, the emergence of content. In order to explain high-level cognition, we need to take into account, as I suggested from the start, the whole, culturally constituted practical context of human experience (see Chapter 2). High-level cognition needs to be explained in terms of structural couplings between the organism and the sociocultural environment:

> Simply put: different resources, different explanatory prospects. Thus the problem of content is only hard—impossibly hard—for naturalists who limit themselves by using overly narrow resources when trying to deal with it. RECers (Radical Enactivists about Cognition) avoid the HPC (Hard Problem of Content) by making appeal to a new set of expanded explanatory resources. In line with a neo-pragmatist tradition, RECers maintain that "the primary bearers of content are semantically articulated symbols, occurring in appropriate dynamic patterns" (Haugeland 1990, 412). The job is then to seek to explain how contentful states of mind actually come into being through a process of mastering special kinds of sociocultural practices (see Clapin 2002, pp. 17–18; Haugeland 1998). (Hutto and Myin 2017, p. 124)

Here, I would like to be specific about how I position myself regarding the question of the origin of meaning, and relatively to this appeal Hutto and Myin make to expanded explanatory resources. By doing so, I shall summarize the central idea I developed in Chapters 2 to 5, and according to which I intend to explain experiential imagination in the following section. As concerns meaning, I endorse a Simondonian and enactive perspective, the very perspective I developed in Chapter 3. In this sense, I take imaginative experiences to be meaningful to the only extent that they consist in embodied and relational, or transactional, lived experiences. I will explain this in the following pages, in the light of Chapter 3 and through the example I introduced in Chapter 4, that of cello bowing. For sure, those two notions, transaction and relation, refer to different philosophical traditions, respectively the pragmatist and Simondon's. But, as I showed in Part Two, they both resonate with the enactive perspective I endorse, especially with the central idea Hutto and Myin discuss in this text, that cognition and meaning arise through a dynamic relation between organisms and the environment, where the environment, in particular the sociocultural environment, plays a crucial role.

But—and this is where I think we need to be radical when talking about the relational or transactional nature of cognitive and imaginative phenomena—I maintain the idea that meaning and contentful states of mind are not products of an organism-world relation. They are not states of an organism. They are not states in an organism's brain. As I highlighted in Chapter 5 (Section 2), cognition is not an internal and representational process enabled and constituted through the organism's engagement with the material world. The imaginative schematism, as I explained it in the light of Malafouris' theory of material engagement, is neither internal nor representational. Rather, it is external and relational. To put it more clearly, as I emphasized in Chapter 2 in Dewey's language (see Chapter 2, Section 1.2), no question for me to consider the mind, taken with its meaningful, contentful and imaginative states, acts and capacities, as an independent, ultimately organic, entity, emerging at the end of a transactional history. The organism-environment relation does not produce or create a mind. Instead, the mind, as Simondon would say, *is the relation itself*. The stake is to explain cognition, in our case, the possibility and effectivity of representational and imaginative acts and capacities, in terms of this individual-world relation. And it is to insist on the technical dimension of this relation, on how technics shapes these imaginative acts and capacities.

What if we located the possibility and effectivity of representational and imaginative acts and capacities, not in physiological processes like brain activations, neither in a hypothetical mind existing somewhere, filled with mental representations and working independently from bodily and technological engagement, in a computational manner—but in a whole practical system spanning over brain, body, and world?

To such a theoretical project, representationalists and internalists use to answer by emphasizing the abstract nature of imaginative acts. "One might suggest, Dokic and Arcangeli say, that motor imagery involves re-creating a proprioceptive experience of the appropriate bodily movements. However, such imagining does not entail re-creating an agentive experience, even if it may accompany the latter" (Dokic and Arcangeli 2015, p. 4). A cellist generally imagines playing the cello when having no cello in hands! As put by Goldman, we can imagine surfing without surfing (Goldman 2006a, p. 47; see also Langland-Hassan 2015, p. 7). In this sense, imagination is a function of irrealization, a way to escape from reality and concrete action (Gaut 2003, 2010; Stokes 2014). Speaking the language of those conceptual oppositions we want to get rid of, imagination is on the side of representation and has nothing to do with concrete action. In the best case, it is a simulation of action, where simulation means neuronal activation with representational effect, without concrete execution. Imagination is "internal" to the brain and to the subject. It is not external; it does not occur in the world of objects, and in the form of a concrete action and of concrete bodily or instrumental engagements. The imagining subject represents things that are absent from the world of objects, in the form of supposedly representational brain activation patterns. The imagining subject is nothing but a representational brain. "It," understand the brain-imagining-subject, imagines privately, independently from any kind of concrete engagement and thanks to its intrinsic representational capacities. "I," understand, me-my-brain, can imagine flying, or playing this Waldstein Sonata which "I" (taken with my fingers) do not know how to play in real life. In imagination, everything becomes possible, independently from the very laws and constraints constitutive of concrete experience, of bodily and material engagement.

At first glance, indeed, composers are able to hear "in their head" very precisely what they want to obtain, in rhythmic, harmonic, and melodic terms. Afterward and by means of instrumental and technical mediations, they work hard to realize what they "have in mind." Classical composers put their ideas on paper before rehearsing them with instrumentalists. An electronic music producer spends hours in her sound banks, tirelessly scrolling and looking for a sound matching as much as possible with her musical ideas or feelings, a sound she will design afterward by means of technological devices and softwares. The ability musicians have to mentally compose might be acquired through practice, by means of technical/technological mediations: playing an instrument, learning composition technics, using a pen and paper, or software for computer-assisted music composition. But, again, at first sight, in the very moment she imagines and creates, in the intimacy of her closed eyes, the composer does not engage in any kind of concrete, bodily and technologically constituted activity. Her imagination seems detached, decoupled from the world, disembodied.

In short, all goes as if this abstract nature of imaginative phenomena made of imagination is the pinnacle of representational cognition. Some even contend that 5EA-style explanations are reserved for cases of "lower-level" cognition (Brooks 1991; Clark 1997), that is, online sensorimotor engagement with the world. In the case of higher level, "representation-hungry" cognitive acts (reading, conceptualizing, imagining, or mentally composing a musical piece, for example), 5EA-style explanations need to be combined with representationalists accounts.

2. Answering the Objection: The Radically Embodied and Situated Nature of Musical Experiential Imagination

Cognitive scientists, whether we talk about representationalists or anti-representationalists, agree on this, that perceiving, executing an action, and having such or such feelings on the one hand, and their relative experiential imaginings on the other hand, share part of their neurological and functional properties (see Goldman 2006a, p. 47; Langland-Hassan 2015, p. 7; Dokic and Arcangeli 2015; Gallagher and Rucińska 2020). A considerable body of research focuses on musical imagination in terms of internal (in the brain) processes. Research employing functional magnetic resonance imaging (fMRI) and magnetoencephalography (MEG) show that musical imagination and musical perception activate similar regions in the auditory cortex (Zatorre, Evans and Meyer 1994). When musicians imagine performing, activity is also found in the same premotor and supplementary motor areas that are activated during performance (Zatorre 1999; Lotze et al. 2003; Zatorre and Halpern 2005). Zatorre, Chen, and Penhune (2007) have shown that neural activity within regions of the secondary auditory cortex can occur in the absence of sounds and mediate the phenomenal experience of imagining music.

Here, the representationalist asks the nonrepresentationalist why one should deny that neuronal activations correspond to the phenomenal experiences of hearing such or such nonexisting sounds, with their intrinsic and characteristic properties, so that those neural activations are representations of those sounds. After all, "representation" basically means making present to the mind something absent. And when specific areas of the brain activate, private imaginative experiences occur. That's a fact. And that must be a good reason to think that the former realizes the latter.

For his part, the nonrepresentationalist takes this reasoning to be fallacious. He wonders, in a Wittgensteinian verve (see Chapter 2), why a necessary condition (brain activations) should become a sufficient condition. More specifically, why should the brain structures involved in having bodily (sensory, motor, muscular, proprioceptive, and affective) experiences be representations of such experiences? Is there not some contradiction in the idea that given brain processes are, on the one hand, physiological parts of a whole human (bodily and situated) relational and practical experience and on the other hand, and at the same time, entities realizing the whole mental representation of this experience? How to justify this functional duplication—and do we need this duplication?—this mereological move whereby what is attributable to the whole system of human experience becomes an intrinsic property of one of its constituting and purely cerebral part? For what reason should we put aside the role of bodily and external factors in the constitution of abstract imaginative experience?

2.1. Experiential Imagination Radically Embodied and the Question of Subjectivity

As I mentioned above, we know that motor imagery on the one hand (taken to be constitutive of experiential imagination by contemporary theorists of experiential imagination) and actual movements, perception, and motor control on the other hand share the same neuronal substrate (Jeannerod 1997; Guillot and Collet 2008; Lotze and Halsband 2006; Moran and O'Shea 2020; Munzert, Lorey, and Zentgraf 2009). Motor imagery involving my hands and arms, for example, activates hands and arms areas of the contralateral motor cortex (Ehrson, Geyer, and Naito 2003). Even more, when imagining, there is covert or residual muscle activity, because inhibition of efferent signals from the primary motor cortex is incomplete (Jeannerod 1997). Engaging in motor imagery of bending one's arm, for example, produces small contractions of arm muscles—measured by electromyograph (see Jacobson 1930, 1932; Guillot et al. 2012). Ultimately, as put by Gallagher and Rucińska, "the pattern of concentric, isometric, and eccentric contractions generated by MI (motor imagery) mirror the configuration during actual movement. And if you imagine that the action takes more effort, there is an increase in the EMG" (Gallagher and Rucińska 2021).

Put simply, an experiential imaginative experience always combines neuronal activations in the motor cortex and minimal bodily activity. Both are inseparable. And the former would have no intrinsic meaning independently from the latter. Minimal bodily activity (muscular contractions, affective states, and so on), constitute experiential imaginings. There is not, on the one hand, a representation or a meaningful image or representation of the saraband in my brain, decoupled from my body, from the kind of radically embodied—muscular, proprioceptive, and affective—processes that constitute my experience of this saraband. And there is not, on the other hand, appropriate bodily movements and affects in which I engage in order to instantiate or realize afterward, in the real world, the mental image I have of this saraband. This would be a classical, representationalist account of the situation. Rather, the imaginative experience I have of this saraband intrinsically is a kind of bodily experience and research:

> We propose that an enactive account of imagination involves a particular type of know-how that is accomplished in the motoric aspects of the rehearsal. It involves not just neuronal activations, but also bodily processes, kinaesthetic and affective processes. (Gallagher and Rucińska 2021, p. 14)

Say that I imagine playing Bach's Saraband in D minor, from the second of his six famous solo suites for the cello (BWV 1008). This is a case of experiential imagination. I see in first person myself playing the cello. I feel in imagination what it is like to hold my bow, pressuring and caressing my four strings with it. I can figure the auditory and proprioceptive difference between playing on metal and on gut strings. I can imagine the difference between playing with different types of rosins on the hairbow. First-person and third-person imaginative experiences merge, combine, and give me access to the whole situation in its globality. More specifically, I have a minimal proprioceptive and muscular experience of what it feels like to start playing a D, doubled in unison on both G and D strings. I am not doing a simple D on open D string. The pressure, and the tension in my left arm, hand, and fingers pressuring the G string, how strings resist, the pulp of the sound, and the way my all body reacts to how the D and the thicker G strings behave together: all is different. It does take more effort. And the physical feeling—something that exceeds mere neuronal activations and consists in a radically embodied, affective, proprioceptive, and muscular experience—of how much effort it takes is intrinsically part of my imaginative experience, and clearly constitutes its intrinsic meaning. This is a feeling of how much effort it takes, shapes, or polarizes the course of my mental imaginative experience. Put differently, what I know about this opening dotted eight D, doubled on D and G strings, has nothing to do with a disembodied "mental representation," preexisting somewhere in my brain. Rather, it is a bodily, muscular, proprioceptive, and affective experience per se.

This, as I just suggested, has to do with knowledge and its radically embodied nature. What does the instrumentalist's knowledge consist in? How does he know the difference between playing a nondoubled and a doubled D? He simply knows how much effort it takes. He feels, in the form of affects, proprioceptions and minimal muscular contractions, what it is like to play this doubled opening

D. As put by Gallagher and Rucińska, who combine the neuroscience of motor imagery with Sartre's theory of analogon (Gallagher and Rucińska 2021), "movement can support and even constitute imaginings." Movements capture something of what this saraband means, of what I understand by this saraband, of what this saraband, in particular its opening doubled D, as a musical experience, consists of.

Reviewing the literature from Descartes through early twentieth-century psychology, Sartre concludes that most thinkers fabricate a naïve metaphysics of the image. As I explained in Chapter 2, this metaphysics consists in making of images pseudo entities existing in the mind and copying or representing existing as well as nonexisting things. Against this, in a way close to Wittgenstein and pragmatists (see Chapter 1), Sartre argues that images are no representational things to be found in the head. Images are not constituting elements of consciousness, as if having a conscious (or unconscious) imaginative experience consisted in manipulating such entities—and as if the mere concept of "neuronal activation" was enough to explain the imaginative life. Rather, appropriating Husserl's language of intentional constitution, Sartre defines imaginative experiences in terms of embodied—and I will add, in the next subsection, technological—intentional constitution.

In Sartre's phenomenological perspective, imagination cannot be reduced to the internal sedimentation of past and recurrent sensations, as is the case in the empiricist tradition. Images are more than the physiological traces of past physical events. More specifically, what Sartre criticizes is the metaphysical framework in which this reduction of images to things or entities in the mind takes roots. According to Sartre, this reduction, proper to the empiricist psychology, rests on a wrong postulate, on the naïve assumption that things in the world carry with themselves their own meaning, independently from any subjective constitution:

> Thus there is no longer any point in distinguishing with Descartes between the thing and the image of the thing, in order to seek afterwards to understand how a relation is established between these two existences. (Sartre 1936/2012, p. 31)

In Kant, against empiricism, the relation between objects of the world and their representation goes from the transcendental subject to the thing. To use a metaphor, objects are illuminated by the subject. The subject represents the world, accesses its meaning, to the only extent that it (the subject) constitutes objects of thought. In this context, the stake is to explain this constitution, to explain how the transcendental subject constitutes or gives to itself an object of thought, makes of a thing the object of a meaningful experience. Conversely, Sartre says, for empiricists like Hume for instance, but also for French philosopher Henri Bergson, the light or luminosity goes from the thing to the subject:

> There is pure light, phosphorescence . . .. There is a sort of inversion of the classical comparison: instead of consciousness being a light that goes from the subject to the thing, it is a luminosity that goes from the thing to the subject. (Sartre 1936/2012, p. 109)

In this second, naïve, perspective, the thing is represented insofar as it is presented. But, retorts Sartre, thinking of the relation between images and representations this way leaves unanswered the question of the adequacy between the image or representation on the one hand and the thing on the other hand. How to explain the adequacy between the image or private representation I have of a thing, and this thing which I am said to represent in my mind, or to imagine? For one does not see how the mere materiality of the external world could by itself provide the conditions of this adequacy, of this correspondence between the thing and the image or representation.

For his part, Sartre assumes that "mental images," as we call them, do not exist, and cannot be found in the head. Rather, images refer to ongoing and embodied processes of constitution. Subjects enact through minimal and covert bodily processes, experiences of things as imagined and meaningful. And the only meaning of these imaginings lies in their bodily (sensory, affective, proprioceptive, physiological, and muscular) aspects. The mind is not a box filled with images, realized somewhere in the form of neuronal activation patterns. Instead, as put by Sartre, imagining my fiancée does not consist in contemplating an image of her, existing as such in my mind. My knowledge of her is not something like an image separated from the bodily movements and affects that constitute the kind of concrete experience I can have of her. Rather, my knowledge of her directly consists of those movements and affects. Radically embodied experience supports and constitutes imaginative experiences. Instead of being a consciousness imagining my fiancée, I constitute, as an affective and motoric body, the experience of my fiancée as imagined. Put differently, as a consciousness, I am nothing more, nothing less than a living and world-directed body:

> But can one distinguish knowledge and movement? In fact, it is not that there is, on the one hand, a directing knowledge and, on the other hand, a series of movements that obey it. Rather, just as one very often discovers one's thought

> by saying it, in the same way one discovers one's knowledge by acting it. (Sartre 1936/2012, p. 34)

In embodied and enactive words, experiential imagination is better understood in terms of an embodied and interactive exploration of the world. It is "something that exceeds the concept of an image appearing before the mind's eye" (Medina 2013, p. 319). My fiancée and my cello are dimensions of my world, which I explore, which I know as a lived and world directed body. And the very bodily processes—movements, affects, and so on—constitutive of my perceptive and agentive engagement with them, are the only processes that support and constitute my imaginative experiences of them.

Great violinist Maxim Vengerov corroborates this embodied conception of musical imagination when he says that music, as it is imagined, written, and performed by musicians, primitively has to do with muscular and affective phenomena. As he puts it, "playing the violin is contraction and relaxation of muscles. That's how it is. And that's how music is written. Depending on whether this is Western music, or French music, Chinese music, Japanese music, these are different kinds of phrasing and different breathing" (Masterclass at the Menuhin competition, 2018).Music is not in the notes but in the bodily moves and affects that produce the notes as coordinated expressive gestures: silences, hand and vocal articulations, breathings, hand position changing, contractions, relaxations, weight, velocity, attack, sustain, and so on. Vengerov insists: "music is between notes. It is not just a romantic saying. It is between notes, how you connect them. Because if there is no connection what we will have is this chain of series of notes that have no connection with each other." Say I play the fourth prelude from Johann Sebastian Bach's cello suites (BWV1010 in E-flat). The time I take to change my left-hand position to reach the upper 5E flat, as well as the time I need to switch from the C string to the A string—each of my gesture, as it is constrained and afforded by a given bodily and instrumental situation, is the concrete instantiation of a musical expressive act, in which lies what I know, as a cellist, about this beginning of the fourth prelude. Let me insist, the very time and movements involved in, for example, a left-hand position switching, intrinsically are musical and musically meaningful, and constitute the cellist's experiential and musical knowledge.

This, as it happens, partly helps thinking of the subjective dimension of experiential imagination in a way different from Dokic and Arcangeli. Imagination is experiential they say (2015), to the extent that it is subjective, that we imagine from the inside a given situation, and feel what it would be like to be in this situation. Agreed. But how to explain it? How to understand that, when imagining playing the saraband, I experience it as "my" experience? As I explained above, Dokic and Arcangeli answer this question by identifying interiority with subjectivity, and by attributing to proprioceptive "images" the only explanative role. But, is subjectivity as interiority anything else than a commonplace metaphor? And are proprioceptions enough to account for subjectivity? Is this equivalence between proprioceptions and the feeling of experience belonging, anything else than a language facility?

What I want to suggest here—in the wake of what I said in Chapter 3, Section 2—with enactivists like Medina (2013), Hutto (2006) and Thompson (2007), is that subjectivity has more to do with a whole bodily and explorative experience, than with mere proprioceptions and motor imagery (see Chapter 3). Beyond proprioceptions, subjectivity and the feeling of self also have to do with worldly engagement:

> The only way to understand "what-it-is-like" to have an experience is to actually undergo it or re-imagine undergoing it. Gaining insight into the phenomenal character of particular kinds of experience requires *practical* engagements, not theoretical insights. The kind of understanding "what-it-is-like" to have such and such an experience requires responding in a way that is enactive, on-line and embodied or, alternatively, in a way that is re-enactive, off-line and imaginative—and still embodied. It involves undergoing and/or imagining experiences both of acting and of being acted upon. (Hutto 2006, p. 52; emphasis added)

The knowledge I have of Bach's saraband is, then, not a disembodied representation that causes in me affective and emotive states, a representation existing as such in the form of intrinsically representational neuronal activation patterns. Instead, the very knowledge I have of this Saraband is a matter of movements and affects, of bodily and instrumentally mediated movements. The minimal muscular and affective events occurring covertly when I engage in the imaginative rehearsing of playing this saraband, provide the meaning and knowledge, the only matter of my phenomenal and imaginative experience. In short, "these are the elements that accomplish the imagining: movement, affect and knowledge" (Gallagher and Rucińska 2021 p. 9).

2.2. Experiential Imagination and Creativity Situated

2.2.1. Online Cognition Situated

Now that I have emphasized the embodied dimension of imagination, let me say why imagination cannot be reduced to embodiment. In what follows, I intend to emphasize the situated nature of experiential imagination in the case of musical and instrumental performance. I will proceed in two steps. First, drawing from Kiverstein and Rietveld's stimulating answer to the "representation hungry challenge" (2018), I will argue that abstract, offline cases of experiential imagination, in the context of musical practice, can be better understood in enactive and situated, instead of representational terms. Second, I will focus on the creative or inventive dimension of cellists' experiential imagination. Cellists' experiential imagination is more than a simple memory of past performances. Cellists are able to change, in imagination, their instrumental performance. In the form of experiential imaginings, they virtually interpret a given musical work in unprecedented ways, representing or explicating to themselves, in the form of embodied imaginative experiences, how they would play a given passage. This is an essential part of musicians' education and preparation (on the mental rehearsing in the context of musical practice, see among others Theiler and Lippman 1995; Cahn 2008; Bernardi et al. 2013). I will define this inventive process in enactive terms and show that instead of being a process internal to the brain, creativity lies in how we respond to affordances that are enacted through temporally extended and coordinative processes involving the body and the social and technical world.

The so-called "representation hungry challenge" (RHC) is to explain how someone could entertain thoughts about nonpresent or nonexisting things, without making use of internal representational states standing in their mind or brain for those things in their absence. How could a cellist imagine playing the cello when having no cello and bow in hand, if not by use and manipulation of mental representations? How could a dancer imagine dancing in a theater while lying in her cold bath, if nothing in her brain functioned as a surrogate for this theater (Clark and Grush 1999; Haugeland 1991; Ramsey 2007)? According to classical internalism and representationalism, imagination, memory, planning, anticipating, among other cognitive acts, need to be explained in terms of mental representations. Such a view, as Kiverstein and Rietveld say, is as widely shared as it is strikingly intuitive:

> Cognitive processes such as imagination, memory and abstract thought might intuitively strike one as being representation-hungry. It seems just obvious that the only way these processes could possibly work is through the mediation of internal representational states. It is natural to think that whenever I think about an object or property x that is absent, counterfactual or abstract, I can do so only by occupying a state that has the function of standing in for x. After all, the thing in the world my thoughts target is not there, and might even never exist. How

> else could I entertain thoughts about that thing if not by means of having an internal state that functions as a stand-in for that thing? (Kiverstein and Rietveld 2018, p. 148)

Enactivists take this intuition to be fallacious. The brain, as Kiverstein and Rietveld (2018) emphasize, does not represent the external world. Instead, it couples affective, motor, and sensory activity. It generates and maintains what Varela calls "micro-identities" (Varela 1999). Micro-identities are recurrent patterns of affective and sensorimotor engagement with the world. These patterns are enacted through concrete practical engagement in everyday situations. A cellist is made up of multiple and mutually nourishing micro-identities, specific to the kind of concrete practical situations she usually engages. She can play as a continuist (playing the leading bass in baroque sonatas, for example), as a soloist, as an accompanist, as the member of a chamber or symphonic orchestra, alone in her bedroom as well. She also can play in a baroque, modern, or jazz style, and so on. Each micro-identity constitutes a unique way to be a cellist, a way to behave in, and adapt to, a specific lived situation.

Varela describes each micro-identity as "readiness for action proper to . . . a specific lived situation" (Varela 1999, p. 9; quoted in Kiverstein and Rietveld 2018, p. 153). Affective and sensory-motor patterns dispose the agent to action, prepare concrete practical experience. Say our cellist plays the continuo part of a eighteenth-century French Sonata for solo traverso and continuo. She positions herself on stage in such a way as she can see and coordinate with both the harpsichordist and the soloist at the same time. She places her hands, body, and cello in the right position, engages her right arm above the strings. By doing so, she engages with action-possibilities (affordances), that of caressing the strings, playing the notes, and coordinating her breath, velocity, and amplitude, and moves with those of the harpsichordist and the flute player. She bodily and actively anticipates and prepares her performance. As Evan Thompson puts it, she engages with the "virtual" (Thompson 2007, p. 74), "that which is real but not actualized" (Di Paolo, Buhrmann, and Barandiaran 2017, p. 228, quoted in Kiverstein and Rietveld 2018, p. 151).

Let me insist, the performance exists virtually without being performed or actualized, in the sense that the cellist relates to it before actually performing it, by means of affective and sensory-motor patterns. Performing is not qualitatively different from being disposed to perform. This, as a matter of fact, is a well-known principle in musical interpretation (as well as in playacting). Playing the first notes of a musical piece always is the most difficult. The interpreter needs to leave the inertia of muscular inactivity in order to play from the start with the same energy she will play the whole musical phrase. To this end, she needs to put herself in a virtual performance state, in a state "as if" she was already performing the melodic phrase. Concretely, she engages in preparative postures and gestures that are intrinsically part of the performance itself. She engages in the very affective and sensory-motor patterns that have been enacted and individualized through past performances—these very patterns of which progress and full actualization realize or instantiate the performance itself. Put simply, interaction patterns enable the cellist to relate to what is not yet actualized, to the distant and the virtual. This, as I will show in due course, is crucial to understand how abstract imaginative acts can be accounted for in non-representational terms.

For now, let me insist on this, that micro-identities form perspectives on agents' environments, depending on which agents relate to things as being valuable and meaningful. Selectively open and differentially sensitive to her specific musical and instrumental environment, the continuist adapts to signals she would ignore if playing as a soloist, over a thirty-musician orchestra, for example. She knows that to eighteenth intimist French music correspond specific bow articulations, that accompanying a traverso demands playing the cello with less pressure on the strings, with a more resonant than penetrating sound, in order not to cover the somehow diffuse sound of the traverso. She knows when and how the traverso player needs to breathe and adapts the way she plays accordingly. In Varela's words, to the cellist's micro-identities correspond specific "micro-worlds," that is, sets of recurrent individual-world interaction patterns, proper to the concrete practical situations she engages.

As Kiverstein and Rietveld emphasize (2018 p. 153), the utmost attention needs to be paid to Varela's notion of "lived situation." Situations are lived not only in the sense that agents actively engage in them. For sure, agents' activities structure the environment, enact specific micro-worlds. But—and this is Kiverstein and Rietveld's ecological touch—the environment also structures agents' activities. "Affordances" (Gibson 1979) refer to action-possibilities, to what the environment concretely provides agents with, in terms of action-possibilities. Playing in a chamber orchestra, for instance, consists of multiple, interrelated activities, that have to do with concrete organizational, orchestral, and architectural arrangements, as well as with instrumental specificities. When entering the rehearsal hall, our cellist readies herself to action-possibilities: sitting on her seat, in the cello space, tuning her instrument after the first cello has finished tuning with the first violin, putting rosin on her bow, her glasses on her nose, getting her pen out, positioning herself so she can see the conductor's hands and eyes—as many stable patterns of activity and social interaction that have been established over time through everyday practice.

These activities are not just realized thanks to neuronal and bodily patterns. They are also determined by the environment, by the kind of action-possibilities (affordances) the environment affords. In this sense, the operations that constitute the mind, that constitute the cellist's knowledge and meaningful relation to her world, are operations of the whole individual-world system. These operations are not just operations in the cellist's brain. Rather, as I emphasized in a somehow provocative way in Chapter 5 when presenting Malafouris' externalist conception of the imaginative schematism, brain operations need to be understood as "extensions of the environment." The idea is that the brain is not a center from which perspective the world gets a sense. Instead, the brain and the world in which we bodily engage, form a set of both relational or interactional patterns. Through their interaction, patterns of the brain, patterns of the body, and patterns of the world dynamically interpenetrate and constitute each other, as inseparable parts. As Simondon would say, their relation is transductive, which means that the brain, the body, and the world do not pre-exist to their relation. They constitute each other. As Kiverstein and Rietveld put it:

> We should not think of the organizational closure at the level of an individual's activities just as the product of recurrent patterns of neural activity taken in isolation from the rest of the body and the settings in which our activities regularly take place. Instead, we should think of these patterns of neural activity as depending upon agent-environment couplings (Bruineberg and Rietveld 2014; c.f. Di Paolo, Buhrmann, and Barandiaran 2017, p. 152). Everyday activities form networks of self-sustaining processes that have among their component parts neural, bodily and environmental elements. (Kiverstein and Rietveld 2018, pp. 153–4)

In one word, the cellist's knowledge, here, is ecological in nature. That is to say, interaction patterns are neurally, bodily, and environmentally realized. They are to be explained in terms of a whole, temporally extended, and socially and technologically constituted practical system.

No need for neurons to emulate internal models of things. The cellist does not need her brain to function as a representational device, in isolation from the body and the world. She does not need her brain to construct and manipulate an internal, intrinsically meaningful representation of her actions, or of her bow, of baroque articulations, of breathings, and so on, in order to orientate and perform her action. Rather, she just needs her brain to maintain appropriate interaction and covariation patterns, as well as to participate in a whole, continuously determining practical situation spanning over the body and the concrete, socio-material environment. Her knowledge lies in these interaction patterns, that is, in how her brain couples affectivity, action, and perception, and participates in a whole practical situation—without internal models being functionally necessary.

Allow me to insist. The cellist might have mental visualizations of bows, cellos, colleagues, rehearsing halls, and so on. No doubt that these visualizations are realized by activation patterns in topographically organized areas of the visual cortex (Kosslyn, Ball, and Reiser 2006; see Chapter 1). But why should we think of these visualizations as representations of bows, cellos, and so on? Why should neuronal activations in the cellist's visual cortex should be taken as representations of her bow, that is, means enabling our cellist to relate to her bow *as a bow*, as something meaningful, something that has the meaning of being a bow, of producing the kind of cognitive, interactional, and expressive life bow actually enact? What would this meaning consist of, taken independently from practical, affective, sensory-motor, and ecological experience? Meaning is a practical and ecological phenomenon. Nothing more, nothing less. The meaning of a bow is not something I possess and manipulate in my mind or brain. It is nothing to be localized somewhere in the brain, independently from the world and the way I continuously engage with the world as a cellist. In this sense, we need to replace mental representations, those neuronal integrations, allegedly meaningful by themselves—with interaction patterns that are practical and ecological.

Of course, the cellist sees her bow in the intimacy of her closed eyes, thanks to neural activations in her visual cortex. Why deny that these neuronal activations are representations of the bow? For the same reason that led philosophers like Wittgenstein and Goodman (see Chapter 2), but also—in a way philosophically incompatible with Wittgenstein and pragmatism—Pylyshyn, to deny that sensory images are representations of things. Pylyshyn opposes to Kosslyn's pictorialist conception of mental images the idea that mental images do not represent things by themselves. Images just resemble things. But the function of resemblance must be distinguished from the function of representation (Pylyshyn 1973). Wittgenstein takes the example of a man leaning on a cane when ascending a curved road (Wittgenstein 1953, §139). There is no way to know if this resembling mental visualization shows a man ascending or descending the road, or leaning on the cane to support himself, or mimicking the posture of an old man, and so on. For his part, remaining faithful to his propositional imperialism (which Wittgenstein does not share), Pylyshyn argues that the only way for mental images to represent things—and to be "images of something" properly speaking—is to be completed with some propositional content. An image "of my grand-mother" is not an image of someone looking exactly like my grandmother for the only reason that a propositional content specifies the meaning of my mental visualization. Another example, one could, "conceive of two images of the identical chessboard with one image containing the relation "is attacked by" and the other not containing it" (Pylyshyn 1973). In one case, the image does represent the relation; in the other case, it does not.

Breaking both with Kosslyn's pictorialism and with Pylyshyn's propositional imperialism, Kiverstein and Rietveld's enactive and affordance-based approach leads to answer the question of meaning in nonrepresentational terms. Coming back to our cellist, the meaning of her bow does not lie in a propositional content, a "know that," or in mere cortical activations either. Again, meaning is a practical and ecological phenomenon. It is afforded by the kind of lived situations bows enable cellists to engage. It lies in how the cellist attunes to her bow, that is, readies herself to engage in specific musical and instrumental situations. In short, cognition is relational and dynamical instead of internal and representational.

2.2.2. Abstract Imagination Situated

That being said, can such a reconceptualization of cognition in enactive, ecological, and nonrepresentational terms, help explaining offline imaginative acts? Is it not that reducing cognition to dynamisms coupling agents to the world is paradoxical and irrelevant as a way to deal with cognitive acts that seem to be best characterized as decoupled? Again, we have the picture of imagination as fundamentally detached from the environment, from the here and now. Some resist to 5EA-style explanations. They contend that these explanations are reserved for cases of "lower-level" cognition (Brooks 1991; Clark 1997), that is, online sensorimotor engagement with the world like those I just described. Accordingly, in the case of higher level, that is, representation-hungry cognitive acts (imagining, conceptualizing, and so on), 5EA-style explanations need to be combined with representationalist accounts.

But those who resist the first waves will be swept away by the tsunami.

For their part, Kiverstein and Rietveld answer this classical objection, defending a strict continuity of higher and lower levels of cognition (see also Koukouti and Malafouris 2020a)—without denying important differences between higher and lower varieties of cognition. Higher forms of cognition they say, "are elaborations and gradual complexifications that develop out of lower, non-representational forms of cognition" (Kiverstein and Rietveld 2018, p. 148). Fundamentally, high-level cognition enables agents to engage with the absent, distal, or abstract. And this can be explained in nonrepresentational terms. To use Gibson's metaphor, landscapes have a complex nested structure: "canyons are nested within mountains; trees are nested within canyons; leaves are nested within trees; cells are nested within leaves. There are forms within forms both up and down the scale of size" (Gibson 1979, p. 5, quoted in Kiverstein and Rietveld 2018, p. 154). The same goes for "landscapes of affordances." Kiverstein and Rietveld use this formula to capture the "richness and interrelatedness of the affordances the environment offers" (Kiverstein and Rietveld 2018, p. 154). Affordances are interrelated at multiple spatial and temporal scales. Agents engage with multiple and coordinated affordances. As I argued some lines above, illustrating Rietveld and Kiverstein's approach, agents' patterned states of action-readiness constitute each other, as part of temporally extended and coordinated processes. When the cellist enters the rehearsing hall, she turns her phone off, takes her jacket off, draws her bow. She prepares to a lived situation to come. The day after the first orchestral session, she executes her instrumental routines, rehearses her lines, bowings, and fingerings, memorizes the conductor's indications. She prepares for the next orchestral session. She engages with the absent and the distal. What she does over longer timescales (playing in an orchestra, preparing a concert), or determines, places constraints on what she does over shorter timescales. All the same, when playing a musical phrase, she engages from the first note with how she will perform the conclusion of the phrase, in terms of musical expressivity and instrumental execution:

> We should not think of offline cognition as a distinct type of cognition, but as a more complex form of coordinating nested states of action readiness and activities to multiple relevant affordances. Such a process is complex because of the nesting of the activities and their increasing reach through time. (Kiverstein and Rietveld 2018, p. 157)

Does this positively explain abstract cases of experiential imagination? Say our cellist relaxes in her warm bath and imagines playing Bach's saraband, as a soloist in Carnegie Hall. She will never play as a soloist in Carnegie Hall. Her learning history and professional perspectives tell her the opposite. All she can do is consoling herself in the comforting, somehow sad realm of dreams. Her dream, at the moment, involves experiential imaginings whereby she deliberately enacts a nonlived experience. She brings into a present imaginative experience, something she will never experience. Are we to explain this in the tempting terms of representationalism?

Kiverstein and Rietveld suggest an alternative: explaining this type of imagining in terms of "re-enactment." Re-enactment enables an agent to occupy states that resemble past experiences. Playing the cello alone, whether on stage or in her bedroom, playing Bach's saraband, playing in front of an audience. As many experiences our cellist had over her learning and professional history, which she reenacts when imagining performing Bach's saraband on stage in Carnegie Hall. Re-enactment, here, has nothing to do with a representational process. Instead, it is a way to engage virtually in a lived situation, without going through the motions of actually doing so. One might object that in order to do so the cellist needs to make use of internal representations of her bow and cello, of Carnegie Hall, and so on. This objection I have already answered. No need to make mental imagery a representational process. The cellist relates to these mental visualizations as meaningful not in virtue of an intrinsic representational power of cortical activation patterns but in virtue of agent-world interaction patterns. Fine, but the objection continues: if meaning is ecological, that is, environment-dependent, how to explain the cellist's ability to meaningfully relate to her bow in its absence? To imagine playing the cello with no cello in her hands, it might be logical that the cellist needs "to make use of an internal model of affordances that acts as a stand-in for the absent affordances, guiding the subject's re-enactment in the absence of the real thing (Foglia and Grush 2011)" (Kiverstein and Rietveld 2018, p. 158).

Two answers, at least. First, affordances are environmental. What a bow affords is nothing a brain neither a body alone can emulate. What a bow affords depends, for a substantial part, on humidity, on rosin's chemical composition, on how strings react in real time, and so on. In this sense, there is very little chance that imagining playing the cello in the intimacy of our closed eyes amounts to emulating a complete realistic experience, thanks to exhaustive internal models of affordances. Second, let's say such an internal model of bow's affordances existed. What would it bring to the explanation that we cannot account in nonrepresentational terms? It might be nothing. Indeed, following Gibson's ecological perspective, talking of mental representations of affordances, would amount to duplicating the only explanative tool we need and already have, namely nested, and temporally coordinated patterns of action-readiness that have been enacted through concrete practice:

> A perceptual system that has become sensitized to certain invariants (information) and can extract them from the stimulus flux can also operate

> without the constraint of the stimulus flux (Gibson, 1979, p. 256). (Quoted by Kiverstein and Rietveld 2018, p. 158)

A cellist attunes through time with environmental structures and conditions of action: bows, cellos, orchestras, stages, theaters, musical institutions, and so on. By means of specialized affective and sensory-motor patterns, she can use this skillful attunement in the absence of these concrete environmental structures and conditions. In this sense, re-enactment consists of a nonrepresentational but affective and sensory-motor state of action-readiness, whereby mental visualizations of things occur, without the need for them to be representational. When imagining playing Bach's saraband in Carnegie Hall, our cellist does not manipulate internal models of action-possibilities. Rather and more simply, she readies herself to such an experience. She enters affective and sensory-motor states of action-readiness that would enable her to coordinate to the affordances constitutive of the lived situation she imagines. Her breath and heartbeats accelerate, she concentrates as if she were on stage with her cello, she enters in a specific emotional mood, and so on:

> We dispute however that it is necessary to invoke an internal model of affordances that the subject somehow manipulates to understand how a subject could pretend to coordinate to affordances that are absent. Models carry information about what they represent based on systematic patterns of covariation that hold between the model and whatever it is modelling in the world. Instead of locating such systematic structure-preserving patterns of covariation in the head, we suggest looking for them instead in the relations of covariation that hold between the patterns of action-readiness and affordances available in the landscape. (Kiverstein and Rietveld 2018, p. 159)

Our cellist does not imagine playing as a soloist in Carnegie Hall out of the blue. Again, patterned states of action-readiness are intimately nested within each other and coordinate over different timescales. Experiential imaginings, understand, moments of action-readiness generating nonrepresentational mental imagery, are situated to the extent that they participate in longer timescale and environmentally constituted affordances. Experiential imaginings are isolated in time in appearance only. In our case, playing in Carnegie Hall as a soloist might be a non-realistic phantasma. In terms of her potentialities, our cellist would remain limited to, and evaluated according to, her concrete, present situation: being an orchestra cellist. Still, playing as a soloist in Carnegie Hall is an institutionally and instrumentally constituted action possibility that determines our cellist over a long-time—or, say, a hypothetical-time—period. As an affordance, it is provided by a specific environment structured by interrelated socio-material practices. To this long-time or hypothetical-timescale affordance, relate all the shorter-time affordances our cellist engages with every day, throughout her life. Among them, learning the cello, passing cello exams, developing her professional career in socializing with other musicians, participating in musical summer courses, masterclasses, and projects, integrating orchestras, and so on. Our cellist's imagination is not a way for her to escape from reality, in an allegedly decoupled realm of dreams. Instead, it is a kind of readiness for action-possibilities, a concrete, institutionally, and culturally constituted way to orientate life and action, to make one's potentialities come true. In short, dreams are meant to become true. And nobody has the ontological right to deny that our orchestra cellist might, one day, play Bach's saraband as a soloist in Carnegie Hall.

In this perspective, imagination is not a state in the individual. Rather, it is a state in the world to which we attune and constantly participate in. It is a state of a dynamical, indivisible, and co-determining individual-world system. And the first constituting elements of abstract imaginative experiences are experiences themselves, in their full and too easily ignored embodied and environmental dimensions—irreducible, in this respect, to allegedly representational neuronal activation patterns.

We are therefore committed to the provocative thesis that talking about "abstract" cases of experiential imagination is misleading, because it lays emphasis on something that should not get that much attention, namely abstraction. Let me come back to this central Simondonian move I made in Chapters 3 and 4. Simondon says that imagination is not a function of irrealization but a function of realization. Fundamentally, as a mode of engagement with the world, imagination "initiates an effective activity of realization." And the modality of the imaginary "becomes that of unreality only if the individual is deprived of his access to the conditions of realization" (Simondon 1965–1966, p. 56). Gallagher is in line with this, when he suggests that when imagining humming a tune, "we do what we would do if we were going to hum the tune, but simply stop short of actual humming" (Gallagher 2017, p. 193; see Chapter 3).

In this sense, imagination is abstract only indirectly, if not metaphorically. The kind of imaginative experiences I described above are not abstract, indeed, in the sense that they do not detach us from concrete practical engagement. As Van Dijk and Rietveld remark in another co-authored paper:

> a narrow focus on these experienced moments of "absence" is misleading. We need to be mindful of the reciprocal constitution of the situation across multiple

> timescales that are easily ignored when describing such moments . . .. We need not think of these moments as "representing" something absent or non-existing but can rather think of them as an experience of participating in an ongoing, still indeterminate, process. Rather than "detaching" from the process, imagination is more fruitfully thought of as opening up the participating individuals further to other affordances that the multi-scaled process of making also provides. (Van Dijk and Rietveld 2020, p. 18)

Experiential imagination, as it operates in the cellist's activity for example, radically couples imaginative agents to the world. It intrinsically consists in engaging with a spatially and temporally extended and coordinated world. Coming back to Goldman's widely shared certitude on experiential imagination, let us answer him something simple: no, it is not possible to imagine surfing without surfing (Goldman 2006a, p. 47; see also Langland-Hassan 2015, p. 7). The only way for agents to imagine playing the cello, surfing, and so on is to be engaged with, and coordinate over short-, long-, or hypothetical-timescales, socially and materially constituted affordances.

2.2.3. *Creative Imagination Situated*

As I emphasized at the beginning of this section, cellists' imagination is more than a mere memory of past performances. In imagination, a cellist changes not only fingerings and bowings but also the expressive qualities of her play, how she feels, how the audience reacts, and so on. There is something qualitatively new in a cellist's imagined performances. When a cellist privately imagines an instrumental performance, she does more than just remembering already-known instrumental gestures and bodily postures, and already-experienced emotional states. She sharpens her gestures, strengthens her capacities and action-readiness, and plays on new emotional strings. She finds solutions to technical problems she encounters when performing a given passage, a given left-hand move, for example. She imagines varieties of ways to touch the audience.

Mental practice has been shown to enhance performance in different fields, including athletics (Feltz and Landers 1983), stroke rehabilitation (Zimmermann-Schlatter et al. 2008), and music (Theiler and Lippman 1995; Cahn 2008; Bernardi et al. 2013). It is generally defined as "a technique by which someone with the intent to practice creates a mental representation of a preconceived idea or action in order to enhance performance" (Bernardi et al. 2013, p. 1). In line with the enactive-ecological and nonrepresentational conception I developed in this chapter, I propose to conceive of mental practice in the context of musical and instrumental activity, as a nonrepresentational part and parcel of a situated imaginative and creative process. I will explain cellists' creativity in terms of a situated and temporally extended process whereby new action possibilities and patterns are enacted, and define mental practice in terms of private, still situated and nonrepresentational experiential imaginings this whole creative process involves. Not that I make no distinction between mental practice and experiential imagination. I just suggest it is plausible to think of mental practice as a specialized and delimited kind experiential imagination, as a cognitive technique involving specialized and intended experiential imaginings.

Summarizing what I said in Subsections 2.21 and 2.2.2, the very action patterns constitutive of cellists' experiential imaginings are enacted through the history of practical engagement. What a cellist does when privately imagining playing Bach's saraband, when mentally rehearsing an instrumental performance, is engaging with affordances proper to her specialized practical activity. She anticipates the performance minutely. She readies herself, and engages in the very affective and sensory-motor patterns of which full actualization would make her perform the saraband and coordinate with the relevant set of affordances (see Bruineberg and Rietveld 2014 on expert ice climbing). As I have shown, this process is nonrepresentational, and it is situated. Mental visualizations and auditions participate in it, as fleeting and nonrepresentational sketches of the whole, practical, temporally extended, and meaningful imaginative process.

From here, as Van Dijk and Rietveld (2020) thematize—in a way clearly congruent with Ingold's conception of creativity as "undergone' (see Chapter 4)—the creativity one observes at the level of private imaginative experiences like cellists' mental practice is part of a practical, situated and socio-materially constituted creativity, proper to unfolding processes of making, producing, and performing. In this regard, Van Dijk and Rietveld's methodology, what they call "philosophical ethnography" (2020), seems a relevant way to think of cellists' creativity and mental practice. Here, I would like to walk in their footsteps, and enter in the detail of the practical and socio-materially constituted processes whereby a given musical and instrumental interpretation creatively elaborates. By doing so, I intend to show that mental practice in the context of musical activity, and the creativity cellists demonstrate when mentally practicing, is situated (non-decoupled) and non-representational.

An instrumental performance is something one builds over long periods of time. Bach's suites for solo cello are part of cellists' entire learning history and life. There are many examples of performers recording the same musical work multiple times throughout their life, making their interpretation evolve each time and in so many expressive, technical (fingering, bowing), instrumental (baroque or modern instruments, for example), as well as technological (the technological devices used for recording; see, for example, Glenn Gould, who recorded Bach's Goldberg variations on different recording devices in 1954, 1955, 1959, and 1981) aspects.

Building an instrumental performance involves concrete engagement in socio-material practices, as well as day-to-day mental practice. We can discern at least six scales in this process. First, the scale of testing different instrumental gestures. Second, the larger scale of musical phrases, in which smaller-scale activities participate. Third, the scale of musical periods, as they relate to the broader harmonic path. Fourth, the scale of the whole musical piece. Fifth, the scale of performing the saraband on stage in a near future. Sixth, playing this saraband in ways suitable to share specific emotional and aesthetical experiences with the audience. Say our cellist plays Bach's saraband (BWV 1008), little by little, rehearsing each bowing and fingering as much as needed to automatize them, to make them easily repeatable. She tries different bowings and fingerings, different left-hand position switchings. More specifically, she evaluates if a given gestures facilitates or complicates the execution of the whole sequence of gestures needed to perform a given passage. For example, on the double opening D of the BWV1008, she tries first to play the D on G string in fourth position. But she notices it is more relevant to play the following E in first position on D string, as it is doubled with a bass A on G string in bar chord. Playing E and A in first position on D and G strings is more efficient, gives more power in the left hand and a richer sound. Also, playing the E on D string enables playing the opening melodic sequence D E E D E F on the same D string, which is more logical in terms of sound quality and color, which also enables playing a rich and sounding A on G string in the bass. Then, considering the shorter distance between left-hand second and first positions, our cellist plays the first double D on G string directly in second position instead of in fourth.

Concretely, here, coming back to Van Dijk, Rietveld, and Kiverstein's conception of affordances as nested and interrelated at different timescales, the cellist is responsive to how given small-scale activities (playing a E in fourth or second position) converge or not with larger-scale activities (that of playing the whole opening sequence of the musical phrase). The affordance of playing the saraband's opening sequence in left-hand second position unfolds over time through practical engagement with large-scale affordances. Responding to small-scale affordances (that of switching between relevant left-hand positions) allows larger scales to keep going. Another example, playing a high instead of a low F in the first opening D E E D E F sequence is a concrete way to generate in the audience a specific, somehow brighter feeling, without leaving the harmonic, somehow melancholic color of D minor. The large-scale affordance (playing on stage in front an audience, sharing with this audience a given emotional state) invites smaller-scale affordances:

> Participating in large-scale affordances then consists in coordinating activity such that affordances across timescales are jointly enacted—a process of coordination that increases the determination of these affordances. (Van Dijk and Rietveld 2020, p. 4)

This is how a given instrumental interpretation elaborates through time, without involving representational states. Considering what I demonstrated in Subsection 2.2.2, experiential imagination, as it enables cellists to imagine playing the cello in such or such way and situation, need not be thought as representational. These moments that seem detached, where cellists privately engage in private experiential imaginings, are better understood in terms of nonrepresentational, embodied, and situated processes.

However, how are we to explain that such experiential imaginative experiences are creative? Thinking of cellists' creativity in the context of Van Dijk and Rietveld's approach, there are at least three ways to explain it. First, creativity of cellists' experiential imagination comes directly from the creativity of practical engagement. Again, the creativity one observes at the level of private imaginative is part of a practical, situated, and socio-materially constituted creativity, proper to unfolding processes of making, producing, and performing. Making an art installation, producing a work of art, building one's instrumental performance are processes whereby new action possibilities and patterns are enacted and, with them, unprecedented experiential imaginings. Those embodied and situated processes dispose the agent toward corresponding kinds of imaginative experiences.

Second, following Van Dijk and Rietveld's interesting suggestion, creativity also lies in indeterminacy, that is, in the multitude of relevant and concurrent affordances one is able to engage when participating or engaging in a process of making, producing, or performing:

> When affordances are conceived as possibilities that get determined in actual activity in real life situations, any engagement with affordances can be more or less imaginative depending on the determination achieved already. Such indetermination is amplified by the multiplicity of affordances unfolding concurrently, reciprocally determining each other. When an inviting affordance

> is still early in the process of enactment (it is still largely indeterminate), coordination with this affordance, given the current situation, may be experienced as imaginative. For instance when the architects at the start of the project imagines what an installation might look like. Conversely, the further an activity has unfolded, the more determinacy and convergence across time- scales, the less of an imaginative character engagement with an inviting affordance has. (Van Dijk and Rietveld 2020, p. 17)

Creativity, in this sense, refers to both (1) a situation felt as imaginative, that is, as open to a multitude of action possibilities, and (2) a concrete process of production, making, or performing, whereby affordances get enacted and individualized. A creation, then, appears as the synthesis and gradual transition between two extremes, that of multiple action possibilities and that of concrete action. The cellist's creation is her performance. And this performance, as it is performed or recorded, carves in the stone of time an individualized realm of the possible that has been actualized, at the expense of other potentialities.

Finally, it might be that creativity also lies in how new openings to affordances that have been previously neglected, re-generate indeterminacy in the process of making. These affordances might come from historical reconstruction (using rehistoricized instruments, baroque bows, and cellos), through technological inventions (amplification technologies, computer-assisted music softwares, synthesizers, plugins, and so on), from new social and institutional contexts. For instance, the study of baroque dance became an important part of contemporary baroque cellists' education. Studying baroque dance enables contemporary baroque cellists to deconstruct the principles according to which twentieth-century cellists used to interpret Bach's solo suites for the cello. Tormod Dalen (2016), for example, strikingly combined embodied approaches with his expert practice of the baroque cello and of baroque dance, to show how different contexts of embodied practices enact specific ways to compose, to interpret, and to listen to music. His analysis focuses on how the practice of French-style court dance, widespread in Bach's time, influenced the composer's writing of his six cello suites.[1] As regards Bach's saraband in D minor for the cello (BWV1008), I remember Tormod explaining to me that instead of playing it legato in a serious and melancholic mood, as cellist of the twentieth century used to play it, a baroque dance-based interpretation may consist in playing it with more happiness and with bow articulations specific to baroque dance specificities. In this context of contemporary historical interpretation, opening up imagination to new creative perspectives consists in accessing new affordances, new action, and interaction patterns constitutive of past imaginaries and practices. Today's cellists who intend to interpret the BWV1008, for instance, find new inspiration in reading Scudery's famous description of a danced saraband:

> She advances, and retreats; she goes slowly, she flies;
> And with such perfect timing, in her agility;
> That an amorous ecstasy robs us of speech,
> And this movement renders us immobile.
> The dexterity of her body, appears in this Dance;
> With her measured steps, she marks the regular cadence,
> Without missing that which the sounds mark.
> The Graces and Love dance by her side,
> And throw, while dancing, subtle hooks
> Which capture a thousand hearts, under the steps of the Beauty
> (Scudery 1660, translated by Rose A. Pruimska in Pruimska 2015)

Here, the very affective and sensory-motor aspects, Scudery describes, refer to institutionally constituted affordances, proper to a seventeenth-century French dancer's bodily and embodied imaginative experience. Creativity is not a faculty of individuals, located in their brain. Nor does it consist in an internal computation on internal representations. Instead, it is a practical and ecological process of performance elaboration. As Van Dijk and Rietveld put it, "imagination is part and parcel of a temporal process in which inviting affordances across multiple timescales are constituted over time" (Van Dijk and Rietveld 2020, pp. 16–17).

Conclusion

5E approaches lead to think of imagination in nonembodied and situated terms. They also offer a conceptual framework that help answer the challenge of representationalism and internalism. The key move to understand in which sense representational, imaginative, and creative acts and capacities are possible without postulating the representational function of the brain lies in (1) emphasizing the ecological, constant embodied, and situated nature of imagination and creativity and (2) taking seriously the temporal dimension of this ecological system. There is no divide between low-level and high-level cognition. Both consist of an online cognitive phenomenon that extends through time, over short and long, even hypothetical time periods. Affordances are technically, socially, and institutionally constituted, and imagining consists in engaging with them, through constant embodied and situated engagement. As for experiential imagination, the ability we have to imagine ourselves playing Beethoven's Waldstein sonata on the piano can be explained in such nonrepresentational, enactive, and ecological terms. As I emphasized, conceiving of experiential imagination as a constant embodied and situated phenomenon allows thinking of its subjective dimension without accounting for it without making a philosophically unjustifiable equivalence between subjectivity and interiority. This equivalence is nothing but a metaphor proper to folk psychology, not to serious philosophical language.

7

Imagination Reconsidered

In the previous chapters, I aimed at combining theoretical and ethnographical considerations. I worked toward an integrative reflection, building my conceptual tools in the flesh of how reality is experienced in concrete situations of making, performing, and doing. My first observation was that the relation between imagination and technics has not been duly studied in contemporary cognitive science. And I assumed that a philosophical enquiry on the theoretical and conceptual frameworks of contemporary cognitive science, combined with ethnographical descriptions, could allow us to overcome this limitation. I worked, as a cognitive philosopher, on enlarging the theoretical horizons and conceptual frameworks in the light of which studying imagination, be it in cognitive science, philosophy, or anthropology.

The question now is to know what I would like the research community to remember from this book, and how I think my approach to imagination can or should contribute to this community. In this chapter, I will summarize the key ideas I developed. I will say positively what I take imagination and images to be, and how I think imagination and images should be studied concretely. It is one thing to say, negatively, that imagination is noninternal and nonrepresentational, and that mental images do not exist. But if one agrees with most of the arguments and observations I developed, then, what is imagination, and what are images, as we do experience them?

In this chapter, I will synthesize the definition of imagination I progressively, but in a scattered manner, elaborated in Chapters 2 to 6. And I will put this definition into question. I will try to formulate some due questions the reader might have. Ultimately, I will show that answering these questions demands embracing a philosophical paradigm that is an alternative to the Cartesian paradigm, on which most of the explanations of imagination and cognition in contemporary cognitive science still depend.

1. Imagination as a Biological Mode of Engagement

In Chapter 3, drawing from Merleau-Ponty and Simondon's appropriations of Konrad Lorenz's notion of "imprint" (see Simondon 1965–1966, pp. 33, 93–4, 97), I defined images in terms of relational schemes. Imprints are patterns of relationship between biologically based behaviors and external stimuli. Sensory-motor tendencies and spontaneities attune through the organism's interactions and engagements with the world. And images, defined this way, prepare perception and action. They are, as Simondon puts it, pre- and intra-perceptive. As preperceptive, images allow organisms to virtually explore their perceptive worlds (anticipation). As intra-perceptive, images are the very biological basis of perceptive and agentive engagement. The idea is simple; perception and action are imaginative in the sense that they are transitory moments in the organism's relation to the world, a relation based on images as I just defined them and on inventive schematism. No arbitrary distinction between imagination, perception, and action. No necessity to rearticulate imagination to perception and action afterward. The stake, instead, is to think imagination in its essential relation to perception and action and, by extension, to life, to the organism's engagements in technically constituted environments.

I talk, here, drawing from Simondon's theory of the image cycle, in terms of "images." A gosling, for instance, learns the "image of 'being a parent'" (Simondon 2008, p. 94; see Chapter 3), in terms of a relational scheme (imprint) with its gaggle that requires specific environmental answer signals. One could object, however, that we do not see why relational schemes, as enacted relational operating structures coupling the organism's sensory-motor patterns and environmental patterns, should be taken as "images." Indeed, in order for relational schemes to be images, one needs to say what an image is and in which sense relational schemes correspond to a good definition of images. I will tackle this point in conclusion. The question will be to know, once we deny the existence of mental images, if there is any other option than just reducing images, as we experience them, to mere material images or to the perception or memory of external representations.

For now, let's just say that I defined images in terms of pre-reflexive relational patterns that define the organism's worlds in terms of affordances (action possibilities) and affective values. In this perspective, as I emphasized in Chapters 3 and 6, imagining is engaging, as situated, sensory-motor, affective and expressive bodies, with action possibilities which themselves result from technically mediated inventive processes. Before summarizing the key arguments I developed in Part Two, let me remark, in line with Gallagher's affordance-based conception of imagination (Gallagher 2017), that defining imagination in such terms is not giving up talking about imagination. On the contrary, it amounts to studying imagination against the backdrop of an already-powerful, enactive, and ecological reflection on life and on the relation between life and cognition. The stake, at the end, is to understand what a theory of imagination formulated in an embodied, enactive, and ecological language looks like.

2. A Schematic Basis of Cognition?

I aimed, in a very Simondonian verve, at reapproaching the question of individual-world co-individuation and of signification (the constitution or discovery of meaning, whereby the living being meaningfully engages with the world) in the light of a reflection on imagination and invention, and more fundamentally on the relation between imagination and technics. This ultimately led me to revive a strong philosophical tradition, associated to the notion of "imaginative schematism." According to this tradition, imagination is not just a function of irrealization, whereby the mind disengages from concrete experience. Instead, imagination is part of experience in general (Chapter 4) and the very condition of experience, the central operator of our cognitive relation to the world. I undertook to rethink, through the lens of Simondon, Stiegler, and Malafouris' approaches, and breaking with Kant's transcendentalism and intellectualism—as well as with Lakoff and Johnson's embodied cognitivism (see Chapter 5, Section 2)—the notion of imaginative schematism. This led me to argue that *cognition, as rooted in life, is imaginative* in the specific sense that it is schematic and inventive. To be more specific, the discovery of significations and the constitution of affordances—the very basis of our cognitive relation to the world—as well as the constitution of cognitive capacities, are schematic and inventive.

Let me take stock on this. Schematism, in Kant, refers to the idea that mental images are produced on the basis of imaginative schemes. Schemes are not images in themselves. They are, in Kant's view, representations of a process or a method for providing concepts with their images. Schemes pertain to imagination but precede mental images. They allow the mind to think what is represented in it under the form of sensorial images. In this sense, schematism and imagination are the functional condition of cognitive (in the general and etymological sense of the term: act or process whereby knowledge is acquired) experience. Kant formulated two interrelated questions, the question of the heterogeneity between concepts and sensorial images and the question of the origin of meaning. First question, how are we to understand that two heterogeneous aspects of thought, sensorial images and concepts, combine in such a way that seeing the image of the number five, for example (5, V, IIIII), we are able to think of the concept of this number? What are the mechanisms or processes whereby this adequation between sensorial images and concepts is achieved, whereby a sensorial image becomes a meaningful representation? Second question, where do these processes take place? To put it differently, to what conditional principle should these processes be referred? Kant answers by locating the imaginative schematism in a transcendental and logical subject.

For my part, I accepted to think of schematism as the imaginative and dynamic condition of cognition. I basically define schematism as an imaginative process whereby a given subject acquires or constitutes knowledge (the cellist's embodied and situated knowledge in Chapter 6, for example), whereby one is able to grasp, for example, the meaning of an object, of a situation, of an action. I refuse to think of schematism in terms of purely internal and intrinsically representational processes. Instead, I take schematism to be practical (Chapter 3), material (Chapter 4), technical, and ultimately ecological (Chapters 5and 6). Schematism is practical in the sense that knowledge and affordances are constituted through concrete engagement in practical and problematic situations. As I emphasized in Chapter 3, learning how to use given objects consists not only in using these objects and passively waiting for them to constrain our actions and shape our intentional structures. Rather, we learn how to use objects to the only extent that, by making, individualizing, and manipulating them, we progressively and transductively discover the (technical and social) operative modes through which we solve the problematics that define our present perceptive, emotive, and agentive, in one word, "practical," situation (adaptation as invention and sense-making; see Chapter 3). I have shown, for example, that musical imaginaries, defined as sets of expressive possibilities and affective values embodied in instrumental and technological devices (cellos, virtual instruments, and softwares, for example), individuate through a process of technically mediated invention and adaptation. This process is schematic, according to Simondon, in the sense that it consists in the free and adaptive play of relational and operating schemes (see the definition above) constitutive of natural and technical phenomena.

In Chapter 5, I argued that schematism understood in such enactive and ecological terms is technical. In Stieglerian words, epiphylogenetic structures, the system of tertiary retentions are not just means to exercise already-constituted cognitive capacities. Rather, they contribute to constituting those capacities. The mind does not exist as a transcendental and already-constituted entity, drawing ideas and concepts from the inside. Instead, ideas and concepts rise from consignation technics independently from which the very order of ideas, concepts, and significations could not be achieved. In this sense, the relation between sensitive images (e.g., images of numbers, of circles or triangles) and concepts (mathematical concepts) results from the concrete and practical inscription of who imagines and thinks in a socio-technical, historically situated context. As I emphasized in Chapter 5, the "who," the subject with its specific mind, is always preceded by the "what," by the whole system of tertiary retentions. Schematism, this process whereby sensorial images and concepts combine, is not a hidden and transcendental art anymore. Instead, it is observable in its constitutive technical traces.

Malafouris' enactive and ecological reflection on the notion of imaginative schematism appeared to me a promising way to (1) articulate my Simondonian and Stieglerian treatment of Kant's question of schematism, to embodied, enactive, and ecological approaches, and (2) emphasize, against any temptation to think of schematism and imagination in internalist and representationalist terms, their ecological nature. Malafouris draws from Hutchins' theory of material anchoring in order to explain the mechanisms involved in the material constitution of the human mind. For his part, Hutchins grounds his notion of material anchoring on Lakoff and Johnson's theory of "image-schematic structures." According to this view, image-schematic structures explain, in a very recognizable Kantian way, how concepts, mapped within our sensory-motor system, acquire a sensitive form, such sensitive forms we can understand as pre-reflexive and embodied minds. Without entering here again in the details of this conceptual elaboration (see Chapter 5, Section 2), I will just say again that, breaking with Lakoff and Johnson's embodied cognitivism and with Hutchins' external computationalism, in a very ecological and nonrepresentationalist verve, Malafouris suggests we define schemata in terms of individual-world relational patterns. Schematism, understood as a relational process, infused and diffused into settings of materially and technically constituted practical activity, gives rise to new and emerging conceptual spaces. Schematism enables organisms to enter new cognitive domains and constitutes cognitive capacities.

Now, that said, if one agrees with the idea that imagination, whether abstract or not, always consists in such enactive and situated processes whereby organisms engage with their world, why say that these enactive and ecological processes are schematic and imaginative? Why talk about enactive and ecological processes in terms of imaginative schematism, a notion that historically roots, with Kant, the cognitive life in internal, imagistic, and representational processes? Why not just talk in terms of enactive and ecological engagement processes, of interactional dynamisms enabling specific imaginative capacities? Why talk, like Fauconnier and Turner (1998, see Chapter 5, Section 2), in terms of "image-schematic structures"? Once we abandon internalism and the idea that images and schemes are internal representations, in which sense is talking about individual-world relational patterns, the same as talking about "images" and relational schemes? Is it not that the enactive and ecological language makes the notion of imaginative schematism obsolete, non-necessary?

Furthermore, is it not that defining imagination in terms of ecological and material engagement processes, amounts to reducing imagination to those processes? Does my theoretical elaboration lead to a kind of eliminativism about imagination, to a kind of reduction of imagination to cognition defined in terms of enaction, individual-world couplings, relational schemes, and affordances—or to a misleading confusion between imagination and cognition? Again, why talk about biologically based individual-world couplings in terms of images and imagination, when the mere notion of relational dynamism, for example, could be sufficient? Why define imagination as such a biological and ecological mode of worldly engagement? Is it misleading or significant for any 5E theory of imagination? Ultimately, what do imaginative capacities like pretense, hypothetical reasoning, empathy, enjoyment of fiction, dreaming, making up a story, abstract musical imagination, and so on become in such a perspective? And what does it consist in, concretely, to study these imaginative capacities? At the end, did I even talk about imagination?

3. Perceiving, Doing, and Thinking Imaginatively

To be clear, the notion of schematism was for me a way to explain the mechanisms underlying imagination understood (1) with Dewey as "a quality that animates and pervades all processes of making and observation" (Dewey 1934/2008, p. 272; see Chapter 2, Section 1) and (2) with Simondon as a general, both biological and technical mode of concrete engagement with the world (Chapter 3). Let me synthesize, here, the key arguments I developed in Chapters 5 and 6, in the light of Malafouris' insights about the enactive and cognitively enabling function of Paleolithic images (Malafouris 2007, see Chapter 4, Section 2). The idea is that practical activities, like material image production, which includes drawing, painting, taking and developing photos, making and editing movies, playing an instrument, and so on, are not ways to represent the world, to realize in the material world static representations of pregiven things (a landscape, a given situation out there in the world, a musical text). Rather, they are better to be understood as forms of perceptual learning, as enactive processes whereby new ways of perceiving, acting, and thinking develop. As Malafouris emphasizes, material engagement, as we could explain it in the terms of an ecological theory of schematism and material anchoring, develops new and specific sensibilities, that is, new and specific dispositions to perceive, act, and think.

For my part, I would tend to put things this way: schematism develops "new aesthetics," in the etymological sense of the term: act and disposition to perceive and to feel (*aisthêsis*). And it develops new poetics (*poiêsis*), that is, new and specific dispositions and ways to engage in action and production. In this sense, I would tend to define, with Simondon, with Kupiec 2009 as well, living beings not as auto-poietic systems only but as hetero-poietic systems as well (Simondon, op. cit. 1962/2010, p. 172; Kupiec 2009, p. 68; see also Di Paolo 2021). Simply said in an enactive language, the very ways we engage with the world as living cognitive systems, the kind of things, lines of force, meanings, aspects, dimensions of reality—in brief, the kind of worlds we are disposed to perceive—are enabled through our imaginative and constitutive engagement with technical mediations and environments. Let me explain this.

As I explained in Chapters 5 and 6, such an enactive and ecological explanatory framework helps thinking the relation between imagination and our daily engagement with the world and in technically mediated processes of making, producing, intervening, observing, and so on. In Chapter 6, I prolonged, in the light of Van Dijk and Rietveld's approach to architectural imagination and in the flesh of my first-hand descriptions of a cellist's concrete and imaginative life, Malafouris insights about the ecological and temporally structured nature of schematic, enactive, and relational engagement processes (Malafouris 2013, p. 39). I have shown that imagination is situated and temporally structured. Imagining is engaging with action possibilities over short-, long-, and hypothetical-timescales, in concrete, technically constituted situations. I emphasized that abstract cases of imagination need to be understood as transitory moments, in the course of embodied, enactive, and ecological processes. Finally, I insisted on the idea that an orchestral cellist's dream to play Bach as a soloist in Carnegie Hall is not a way to disengage from reality and escape to a comforting world of fantasies. Rather, it is a way for her to engage with reality and with the realization of herself. As a daydreamer, the cellist enacts herself and her world. This dream, just like any abstract imagining, is situated, constituted through concrete and constant engagement with the world. And as I explained in Chapter 3, these affordances, constitutive of our imaginings, are technically and institutionally constituted through practical and inventive engagement processes. In brief, no one imagines and individuates out of concrete, materially constituted, and problematic situations.

To go a little further here, I would like to show how this explanatory framework helps thinking the relation between imagination and artistical and political engagement. Again, the main idea is that the ways we engage with ourselves and with the world, are imaginative, not just perceptive, emotive, agentive, and intellective. Put differently, in this book I wanted to understand in which sense the ways we engage with the world, whether on stage with a cello or in socially and politically meaningful situations, are imaginative. I needed to think how the very dispositions to perceive, act, and think—aesthetics and poetics—that define us as engaged or disengaged actors in given situations, are imaginative, or, more specifically, are enacted through imaginative and inventive processes of making, of doing, of intervening—these processes taking place in concrete, technically constituted practical context (technical environments, infrastructures, institutions, and so on). Again, imagination is not a faculty. It is a quality and a dynamism, intrinsic to our constant relation to the world. I had to understand and explain that. I wanted to show how organisms, through ecological, materially, and technically constituted processes, imaginatively transform themselves and their enacted worlds.

To illustrate it here, as I was elaborating my reflection in Chapters 5 and 6 through the lens of Malafouris' intuitions, especially in the light of his definition of paleolithic images in terms of perceptual devices whereby organisms develop new and specific sensibilities and dispositions to perceive, act, and think, I made a little movie, on a homeless urban artist.[1] Papillon (Butterfly), as he calls himself, is a young Caribbean homeless, who settled on a little unattractive square in front of Gare de l'Est in Paris. To quote the jury of the festival *Athmosphères*, who kindly mentioned and awarded my little movie, Papillon is "an atypical man who tries to build 'his *dream life*' in the heart of the city, by giving life to his hybrid universe from materials retrieved throughout the city, sublimating the debris rejected by nightmarish lives"[2] (my emphasis).

The starting situation of this little audiovisual production combined different aspects. First, there was the concrete situation of a homeless young man living in the streets, interacting or not with passersby. And to put it simply, like any other passerby, I was passing by with no constituted and assumed ways to act or not to act in this situation, with no constituted ways to think the situation and with no constituted ways to enter the perceptive space of the square. This situation was problematic in a Simondonian sense. From my personal perspective, it was problematic on a perceptive and emotive level. How to feel and behave, in between respectful and disrespectful passivity, in the perceptive space of the square? What to say or not to say, what to do or not to do when entering the scene, a public square, but at the same time, the intimate space of someone whom I do not know, whom I disturb, whom I am watching both out of curiosity and in spite of myself, as he goes about his daily business? It was also problematic on a moral and political level. I was not convinced, for example, that I morally had to help him. Who am I to think I will help people and to think that someone I don't know needs help, and my help—following what easy, miserabilist, and caricatural understanding of the person's situation? Even if he needed some help and if I wanted to help him in some way, I lacked solutions for him, and was unable to change the unfair world with my little hands. I was not part of a neighborhood association helping homeless people. In brief, I did not know exactly what to make of this situation, and how to position myself emotionally, intellectually, and concretely in it. From the point of view of Papillon, the situation was also problematic. To put things quickly here, how to make his place both a public and an intimate place? How to survive? How to get through people's judgment, uneasiness, superficial interest, voyeurism? How to create a social bound when he is categorized as a stranger in the Ancient Greek sense of the term, that is, someone who is not a citizen (he is an undocumented person, he has no rights, no political voice, no social and no economic value, no existing profile on social media)? How to feel at home and keep being motivated when police officers systematically destroy his place and artistical creations?

Second, there was my need to make a movie. And there was Papillon's dream to play the piano, one day, on stage, for an audience he could share his thoughts and emotions with, about life, dreams, freedom, the climate, political oppression, and the cruelty of living in the streets. How to constitute that audience, and realize our respective dreams? I took a camera and started filming. Papillon, who never has the opportunity to speak and to be heard, was enthusiastic. My two cameras, as well as image-editing software, dramaturgic rules and techniques for writing a story, technical constraints in the setting of the scenes, the application of principles for compositing images, and so on changed the situation and offered Papillon and myself new action possibilities, specific ways to relate to each other, to imagine, and to engage in Papillon's place. Our dreams merged without confusing. We started to dream together in different but interrelated ways. The camera structured our relationship and the kind of affordances we imaginatively engaged with. We ended up on stage in a cinema; in the festival *Athmosphères*, my film projected, Papillon and I playing together for an audience sensitized to Papillon's situation and to how I perceived, thought of, and engaged in, this situation. This was an aesthetical and poetical sharing.

I will not get into the detail of this here. This would demand a dedicated chapter. What I can clarify, however, is in which sense this process of film making was imaginative. It was not imaginative because I was making a movie, filming, producing, editing, composing, and arranging external and technically made visual and auditory images (relation of reduction between images and material images, between imagining and engaging in a technical process of making). Rather, this production and realization process was imaginative first in the sense that it was a creation or production process involving imaginative capacities like anticipation, virtual exploration of the scenes, of events, and of technical limitations and requirements, imagining a story, dramatic tension, imagining poetical situations and compositions of images, as well as the structure of my film, its colors, rhythm, beginning, middle, and end, and so on. As many imaginative capacities proper to audiovisual production, some of them specific to nonfiction cinema. These capacities individuated and individualized through technically mediated processes of making, filming, writing, editing, composing, and arranging. Imagining this film as a cineaste was not simply visualizing it independently from the situation, from the square, and from the technical devices I manipulated. Instead, it consisted in being concretely involved in the very situation I, with my camera, micro, computer, and softwares, participated in enacting.

In this respect, my imagination, the kind of imaginings I was disposed to have privately in relation to the situation, was that of a cineaste, not the imagination of a neighborhood resident for example, or of a police officer, or of someone who, instead of making poems on, or with, homeless people, help them as a member of a neighborhood association, walking from square to square. In this regard, coming back to Malafouris's understanding of Paleolithic images as enabling sensibilities and dispositions, what I became, as a cognitive and imaginative system, was enacted through technically constituted engagement processes. And becoming the specific cognitive system I am, disposed to perceive, act, and think, as well as imagine and produce material images in the ways of a cineaste, consisted in imagining in specific, technically constituted ways, through concrete engagement in a practical situation. Schematism refers, here, to these ecological, materially and technically constituted processes whereby, as an embodied and situated agent, I imaginatively enacted and transformed myself and my world—whereby new aesthetics and poetics emerge.

This process was also imaginative in the Simondonian sense I explained above and in Chapter 3, that it was schematic and inventive. I had to discover, to invent, through imaginative and technical processes, the relational schemes, the operative modes allowing me to restore compatibility between the different dimensions of the initial problematic situation. The same way the instrumentalist individualizes his instrument and makes of it a prosthetic and expressive extension of himself, I had, for example, to learn how to hold, manipulate, and configure my two cameras. They now embody an evolutive choreographic grammar through which, as a cineaste and a cameraman, I affectively engage with space, time, situations, and people. Ultimately, it was imaginative also in the sense that from the start, as I emphasized above, making this film, producing it, consisted in having, transforming, merging, and realizing dreams. In one synthetic formulation, imagining was living (imagination as life), as a cognitive being (cognition as life and imagination), through this situation. And finally, this process was imaginative in a qualitative sense. My ways to perceive, act, and think in the situation became imaginative. If, as says Dewey, imagination is a quality of experience, one could say that I was an imaginative and creative cognitive being in the sense that I found original, inventive, and solving ways to perceive, act, and think in this situation.

4. Imagination as a Function of Realization and Enaction

This leads me to emphasize what I take to be the most important idea of this book, that, as a transactional or relational phenomenon, imagination is a (biological and technical) function of realization and enaction, irreducible to a mere function of irrealization and representation. This way of talking about imagination without reducing it to its function of irrealization is not original. Nor is it proper to Simondon or, more largely, to a continental philosophical tradition. Without even talking about enactivists, whom I abundantly mentioned in this book (to name only few of them again, Malafouris 2007; Hutto 2007; Medina 2013; Gallagher 2017), the idea is, for example, well formulated by Richard Moran, who assumes that

> a better understanding of the imagination . . . lies in loosening the grip of a narrow concept of imagination that is exclusively concerned with the representation of

> various states of affairs, that restricts interest in it to the imagining of various fictional truths, and thus suggests an unreal discontinuity between what and how we imagine and our real-life engagements in everyday life. (Moran 1994, p. 76)

For his part, reflecting on the relation on imagination and technology, especially digital technologies, Turner (2020) argues, in line with the theory of images-schematic structures (Fauconnier and Turner 1998; Lakoff and Johnson 1999), that imagination is not to be treated as either related to the production and manipulation of mental imagery or as synonym of creativity. Instead, imagination is the conditional basis of cognition. As put by Johnson (1987, p. 172, quoted in Turner 2020), "imagination is central to human meaning and rationality for the simple reason that what we can experience and cognize as meaningful, and how we can reason about it, are both dependent on structures of imagination that make our experience what it is." In this book, I aimed at emphasizing imagination's essential relation to experience and concrete engagement, in embodied, enactive, and ecological, instead of internalist and representationalist, terms. Also, I argued that it is possible to explain imagination as we usually talk about it in terms of pretense, hypothetical reasoning, empathy, enjoyment of fiction, dreaming, making up a story, musical, cinematographic, choreographic imaginations, and so on, as specific, technically enabled, and individualized cases of imagination understood as such a function of realization and enaction. For sure, I did not properly explain each of these imaginative activities. This would be more the topic of a dedicated book and more largely of a research program.

For now, let me summarize, here, what "imagination as a function of realization" exactly means. In view of what I developed in Chapters 3 to 6, imagination refers to the very relational, embodied, and ecological phenomena whereby individuals and collectives enact their world and themselves as cognitive systems. As such a function of realization and enaction, imagination primarily has to do with action and production, in the sense that producing and doing are part of realization and enaction phenomena. It is not without reason that Simondon concludes *Imagination and Invention* (1965–6) with a reference to Alfred Espinas' praxeology, "science of the most universal forms and the highest principles of action in all living beings" (Espinas 1890). Throughout his lessons on imagination, Simondon does not cease to present imagination as the highest principle of action in all living beings. As anticipation, imagination prepares action. As integration of the data of the sensory-motor experience, which it then systematizes in the form of imaginaries and symbolic universes, imagination makes action the insertion point of the world in the subject and determines the way the subject inserts itself into the world. As invention, imagination concretizes the action and redefines the context of its execution. For their part, enactivists define "enactive imagination" in terms of concrete and enactive engagement in action. As Medina puts it in the light of Wittgenstein's insights, in his later writings, about the agential aspects of imagination, "imagining is a doing, an activity that is not reducible to simply having an image in one's head" (see also Brann 1991, p. 6; Turner 2020). Accordingly, imagining is even "more than an intellectual act and can be a volitional act" (Medina 2013, p. 318). And finally, as I insisted in Chapter 6, imagination defined in enactive and ecological terms is a function of enaction, and imagining is engaging, as situated agents, with action possibilities. The cellist imagines. She dreams. And as a daydreamer she enacts herself and her world. I will not dwell on this again.

Instead, I will draw a somehow provocative—but, as I will show, philosophically relevant—conclusion from it. In this book, I emphasized that engaging with the world as cognitive systems is an imaginative and inventive phenomenon—as much as a perceptive, emotive, and agentive one, as it is generally emphasized by enactivists. I aimed at explaining interactional dynamisms as imaginative, and not only in terms of mechanical interactions processes devoid of imagination.[3] Again, imagination is not a state in the individual, independent from the individual's engagements with the world. It is a state in/of the world, the state of an ecological cognitive system spanning over brains, living bodies, and the socio-material environment. And to this extent, cognition, basically defined in terms of embodied, enactive, and situated engagements with the world, is imaginative, or at least, has some essential relation to imagination. In a strict continuity with what I said just before and developed in Chapter 6, I would say that, defined in terms of embodied, enactive, and situated engagement with the world and with oneself, cognition is a daydream that comes true, that realizes itself. It is not just saying that imagination is to be understood in the terms of an enactive and ecological conception of cognition. Instead, it is saying that cognition needs to be explained in the light of a theory of life as an intrinsically imaginative and inventive phenomenon.

This way of talking might appear provocative, especially in a scientific context historically more used to emphasize the purposedly mechanistic and computational nature of the mind. In any case, as I emphasized in Chapter 6, imagining is engaging with affordances as well as in their constitution through concrete engagement in action and production, across multiple timescales.

Imagining is engaging concretely in action and production. To quote Simondon again (see Chapters 3 and 6), as a mode of engagement with the world, imagination "initiates an effective activity of realization." And the modality of the imaginary "becomes that of unreality only if the individual is deprived of his access to the conditions of realization" (Simondon 1965–1966, p. 56). In this perspective, if dreams do not always become true, it is not because of an arbitrary and unreal distinction between perceptive and agentive, ultimately cognitive experience on the one hand and imaginative experience on the other. And realizing one's dream does not consist, in a hylomorphic manner, in materializing, thanks to action and production techniques and to technical and social mediations, imaginings as if they had nothing to do with realization and enaction processes and with how we cognitively engage with the world and with oneself. Instead, realizing dreams is just having dreams, to the only extent that having dreams is realizing—as well as failing to—realize something. It is enactively engaging with the world and with oneself, over short, long, or hypothetical timescales. Again, as I put it in Chapter 6, functionally speaking, dreams are meant to become true, for the reason that they are part of realization and enaction processes (doing, making, producing, performing, and so on).

This way of thinking invites us, as I emphasized in Chapter 4, to pay particular attention to the essential relation between imagination and cognition. Bachelard beautifully emphasized the primary dreamlike dimension of the mind. In a famous lecture he gave in 1954 on the RTF radio,[4] entitled "awake dreamers," Bachelard argues that "our belonging to the world of images is stronger, more constitutive of our being than our belonging to the world of ideas."[5] In the light of Ingold's appropriation of Bachelard's theory of imagination, I emphasized in Chapter 4 that imagination is the beneath of perceptive and agentive experience. Perception and action root in imagination understood as (1) a biological and general mode of perceptive and agentive engagement, and (2), in the wake of (1), the impulse and imaginary beneath of perceptive and agentive experience. I observed, for instance, that when performing a musical piece, a cellist, more than merely producing sounds with a cello and being constrained by her cello and bow, engages in affective and imaginative experiences. Playing the cello is not just a technical (instrumental) act, it is an imaginative act. The cellist does not just make the strings of her cello vibrate. To take Ingold's poetical metaphor again, playing Bach's Corrente in D minor, the cellist becomes a Pinési, a kind of imaginary bird. "Its body the wind, its breast the tempest, and its wings the sky" (Ingold 2013b, p. 738). This is a simple thing to say, playing the cello has to do—as it is also the case in theatrical or choreographic expression, for example—with the imaginary. On stage, with its body, instruments, and props, the performer enacts something more than herself, something even more than human. She participates in a "more-than-human world" (Ingold 2013b, p. 739, see Chapter 4). She becomes, as a sensory-motor, affective, and prosthetic (coupled with her instrument, see below) body, an imaginary creature, of which constitutive forms and dynamisms are isomorphic to those of the material and natural world. Not that our cellist does *as if* she was a Pinési, trying to resemble its external representations, in a purely figurative and mimetic way. Nor does she imagine being a Pinési in the weak sense that she would consider, without believing it, *that* "she is a Pinési." Rather, she feels in her flesh, a prosthetic flesh, what it is like to be such an imaginary bird. Imagination, did I say, is "experiential" in the sense that it has to do with the experienceable. I will come back to this in conclusion. Here let me just say again that the cellist experiences being something else than a human body. She experiences being on the scale of the sky, as violent as a tempest. And the audience feels carried away in the vibrating and sounding space like leaves and buildings in a tornado (see violinist Itzhak Perlman describing violinist Sacha Heifetz as a tornado, in Bruno Monsaingeon's *The Art of Violin*). Playing alternatively on A and D strings triggers a kind of bodily and imaginative metamorphose, whereby the cellist becomes an embodied imaginary creature or phenomenon. This description, one might object, sounds poetical, but far from being scientific. I offered, however, theoretical tools in the light of which accounting for this (non-only) Bachelardian intuition in the language of cognitive science, especially in the language of 5E cognition theories. This led me to offer a conception of imagination as essentially experiential (Chapter 6; see Section 2 of this chapter), and to reject, with enaction, any divide between perceptive and agentive engagement on the one hand and imagination on the other.

5. Coming Back to Renaissance: Toward and Anthropology of Fantasy and Magic

To reflect on the philosophical meaning of this approach, as I suggested some lines earlier, it might be that this way of thinking of imagination as a dreamlike function of realization and enaction, as an enactive function of dream realization, leads to reviving the very philosophical tradition that has been outshone by the Cartesian and mechanistic conception of images and of the mind. It might be that my enactive and ecological approach to imagination revives, against Descartes' mechanistic conception of images understood as static representations of things, a conception of the mind as fantastic and essentially imaginative, and a conception of imagination and images as magical—a conception defended by the philosophers of the Renaissance (see on this Ansaldi 2013), as well as by Orpheus' children, poets, and musicians.

To begin with, let me quote a dear friend of mine, one of contemporary greatest masters of the baroque violin, Amandine Beyer:

> You're right about the magic. What I often find in music and even more clearly in the sharing moment of the classes, which is a moment where actually everyone is searching together, is that this magic becomes audible. It's like a live alchemy session . . . time to time it works, and I see rays and lightning and sparks coming out of the room. (Amandine Beyer, January 2022, personal correspondence)

Here we find again, in a beautiful musician's language, the same intuition I developed with Ingold and Bachelard just above and in Chapter 4, that perceptive and agentive experience, as it takes place in musical performance, for example, is a hyper-real imaginative experience (see Chapter 4, Section 1). In those moments of musical sharing Amandine describes, she is like Bachelard's Ojibwa boy, with his Thunder bird. But we also find, in Amandine's words, the idea that there is something magical in musical performance and imagination. To give a philosophical account of this and, by extension, clarify the philosophical meaning of the enactive and ecological conception of imagination I developed in this book, let me summarize the historical debate I just referred to, between Descartes's mechanistic conception of images, and the conception of mind as fantastic and of images as magic.

The history of the notion of imagination knew a mechanistic turn, that substituted to the purposedly occult and magical power of images and imagination, as it was conceived by philosophers of the Renaissance, the mechanist metaphor of impression. According to this famous metaphor, imagination is a kind of wax in which the shape of things is imprinted. Images, instead of dynamical constituents of the world combining material and spiritual qualities, are simple duplicates, copies of forms already given out there in the objective world, those images projected in an internal theater, in front of a mind's eye. This way of thinking still underlies contemporary conceptions of imagination in cognitive science. To mention just one example, "this sense of imagining is at work in the platitude that imagining involves image or picture-like mental states. We describe ourselves as *visualizing*, or as *seeing an image in our mind's eye*, or, yes, as *imagining*, where the 'image' in 'imagine' is emphasized" (Langland-Hassan 2020, p. 5). Without dwelling on this more than necessary here, in his *Dioptric* (1637), Descartes appropriates Kepler's optics. In the terms of a mechanistic conception of vision, Descartes considers that images are printed on the retina through refraction processes. Retinal movements are transmitted to the brain and the pineal gland by the nerves. For sure, Descartes's conception of vision is not reducible to this retinal mechanism. According to Descartes, our ability to reach a visual perception of the figure, of the distance, of the depth, as well as to relate intentionally to what we see, depends on what Descartes calls a "natural geometry" (Descartes, *Dioptric*, VI). For Descartes, this natural geometry refers to the active function of imagination, to a psychological process whereby one is able to recreate spatial three-dimensionality from raw visual data received in two dimensions and made of retinal colored and luminous spots (for more details, see Bellis 2020). But still, in this mechanistic perspective, images by themselves are defined in terms of retinal pictures (*pictura*), of static copies of things. And imagination is a function of representation, whereby what is perceived is mentally apprehended according to an objective model.

However, before and as opposed to this mechanistic conception of images devoid of any intrinsic spiritual quality, Giordano Bruno, for example, a Neapolitan philosopher of the Renaissance (1548–1600), takes images to have power, a power of composition and signification that makes them irreducible to mere copies or representations of things. In his *De imaginum, signorum et idearum compositione* (*On the composition of signs, images and idea*s, 1591), Bruno deeply analyzes what he calls the power of imagination or fantasy. And he does so in a way that somehow resonates with my understanding of imagination in terms of an enactive and ecological phenomenon whereby significations, as well as cognitive abilities and dispositions to perceive, act, and think, emerge. Bruno draws a radical distinction between the Eye of God and the eye of the human. God sees everything the same way He sees Himself, because He Himself is everything (1591/2009, p. 486[6]). By contrast, one cannot see oneself seeing, for one cannot "distinguish oneself from oneself[7] (*Ibid.*). God can see Himself as everything in a pure reflexive way, with absolute clarity and transparency. As for us, to see ourselves, we need the mediation of what Bruno calls "mirrors." The notion of mirror, in Bruno, refers to the idea that in order to see ourselves, to know ourselves and nature intellectually, we need the always-biased, incomplete, and polarizing mediation of external compositions of simulacra, of images, of figures, and of signs. Let me quote Ansaldi in his stimulating *Fantastic imagination* (2013):

> The intellect is defined in relation to exteriority more than in relation to interiority. It is the interiority which is contained in the exteriority of the images. Or better: it is determining for Bruno to think the constitution of the interiority of the *mens* (human intellectual faculty) in relation to the exteriority of the vision by images. (Ansaldi 2013, p. 352)

Bruno's conception of cognition (the process whereby knowledge is acquired) or intellection reverses Plato's metaphysics and refutes the Platonic condemnation of simulacra. In Bruno, simulacra, signs, and figures are the constitutive and visible mediations of knowledge (for more details on this, see Ansaldi 2013, p. 374). Translated into the language I used in this book, images, as external representations (signs, figures, signifiers), are perceptive devices whereby organisms enter new conceptual realms, whereby cognitive capacities are enabled, significations, enacted, and knowledge (see Chapter 6 on the cellist's embodied and ecological knowledge), constituted.

To summarize things, there are two main ideas to draw from Bruno's conception of imagination as fantastic, two ideas I take to be absolutely central and which I think my anti-representationalist, enactive, and ecological approach to imagination as technically constituted, aimed at emphasizing as well.

First, in Bruno's perspective, images need not be understood as representations, as copies, but as constitutive parts of signification processes. Bruno argues that "images are not defined because of the things signified, but more according to the condition of the signifiers themselves"[8] (Bruno 1591/2009, p. 526, quoted in Ansaldi 2013, pp. 375–6). Put differently, images have a power, lying at the articulation of the intellect and fantasy (imagination), which is to compose human visibility, to constitute humans' ability to see and understand. Here again imagination becomes part and condition of cognition. To quote Ansaldi again, "the being of the image does not depend at all on the representative power of the *mens*, which has for function to realize the junction between the model and the copy, the 'thing' and the 'simulacrum,' but on its power of figuration and visualization, that is to say on its aptitude to create itself the things signified by the images." Put simply, images are not "copies" of things. And they are images properly said, and meaningful, to the only extent that they constitute mirrors and signified, that is, visible sets of images, figures, and signs enabling us to see and understand.

Second, and in the wake of it, Bruno makes of this power of composition of images, power of fantasy or imagination, the very basis of all cognitive activity. To quote Ansaldi again, "one could say that Bruno elaborates in the

De imaginum compositione a kind of 'anthropology of the fantastic', a theory of the human nature making of the power of the fantasy the neuralgic center of all the other faculties" (Ansaldi 2013, p. 383). In Bruno, images have a power of composition and signification, independently from which the *mens* would not develop. Bruno, then, suggests that it is possible to think of imagination as the basis of cognition (intellection, passion), without reducing the one to the other. I will dwell on this in the next section, in the language of contemporary cognitive science.

For now, and to put it simply, I can say that I would consider my book useful to the extent that it provided theoretical and conceptual tools allowing contemporary cognitive scientists to work in the sense of such an anthropology of the fantasy. But I would also fulfill my goal if I managed to give cognitive scientists theoretical reasons to believe in magic. Against Descartes, and against any temptation to reduce images to mere copies of things, as well as against any temptation to reduce imagination to a mere faculty of representation—and finally, against any temptation to reduce cognition to mere interactional, biological, and environmental dynamisms, or mechanisms or processes devoid of fantasy—images have power, a power of composition and signification. This is the basic definition of magic, of images as magic, and what I take as a primary conclusion of a 5E theory of images and imagination. In view of my approach to imagination, indeed, this power makes of imagination a magical function, in the sense that images, as intrinsic and dreamlike dimensions of realization and enaction processes, have an effect on the world, on the individual-world system.

Philosophically speaking, without dwelling too long on it, this way of thinking of images as magical, as having power, did not cease after Bruno's early, slow, and horrific death. Explicitly against Descartes, in his time, Spinoza emphasized that ideas and images are not just static representations of things. Images and ideas, Spinoza says, are not mute. They speak by themselves (Macherey 1994, p. 332):

> One considers these ideas only as mute figures drawn on a board, and the preoccupation produced by this prejudice prevents from seeing that any idea, as an idea, envelops the affirmation or the negation. (Spinoza, *Ethics*, II, proposition 49, corollary)

Against Descartes and Kant, Bergson made an interesting distinction between schemes and images. A "dynamical scheme," for Bergson (1919), is not a static image but a dynamism of image production and composition, and the very basis of what he calls the "intellectual effort," whereby thought and creation are achieved. Thinking and creating is, for Bergson, the result of a process whereby a pre-reflexive and dynamical scheme, an unclear intuition, develops in clear and distinct images:

> It is for the musician or the poet, a new impression to unfold in sounds and images. For the novelist or the playwright, it is a thesis to be developed in events, a feeling, individual or social, to be materialized in living characters. One works on a scheme of the whole, and the result obtained when one arrives at a distinct image of the elements. (Bergson 1919/2009, p. 175)

Dynamical schematism consists, for Bergson, in developing and realizing images and thought. It refers to a dynamical process, to an effort whereby, beyond representing a pregiven state of the world or mind—the illusion of the possible in Bergson, which he replaces with the virtual—one composes those states. Bergson undoubtedly remains intellectualist as he defines schemes and images in terms of representations, referring those representations to the pure interiority and duration (*durée*) of consciousness. For my part, I explained this power of composition and signification, in terms of realization and enaction, in embodied and ecological terms. But the general idea here is that images, as integrative parts of a schematic dynamism, have power, are movements of development, of progression, as Bergson would say (see Deleuze's *Image-movement* (1983), his analyses and application of Bergson's theory of images and schemes to cinematographic creation). To put it differently, thinking of individuation, realization, and enaction processes was for me a way to emphasize the essential, dreamlike, and imaginative, dimension of cognition. And conversely, it was a way for me to redefine schematism in noninternalist, nonrepresentationalist, and nonintellectualist, but embodied and ecological terms.

If what defines something as magical is its ability or power to affect reality in extraordinary ways, then images have a magical power, a power of polarization and structuration of reality, according to dreams, which a purely computational and rational mind would evaluate as impossible, fanciful, and statistically irrelevant.

The poet knows this magical power of imagination, of musicians, poets, and so on. Let me refer to Paul Valery's verses on the composition and magical power of Orpheus's song. Orpheus puts inert and dispersed stones at work by singing to them. Stones compose a palace:

> I compose my spirit from under the myrtles—Orpheus,
> I write, Exemplar thee! . . . Fire falls in descendent spirals:
> He transforms the bare mountain, which awed is found—
> A mountain cowed, a mountain crowned in majesty;

Whence resonance exhales, pure as the art
In an act of a sacred deity.

If the god should sing, as he sat on the rim
of a sky—t'would echo its own resplendence!
A wail would stun, a cry would shudder the high
harmonious hall—a call would ring, its golden walls would sing;
Return him the song of the sanctuary.

Sing Orpheus did, so to recount this scene:
The sun saw the horror of the moving stones;
A rock stepped; surprised, stumbled then—each flint, bewitched,
Felt dragged to sense, so found the azure-trance, awoken within
The living fire of delirium!

Evening bathes the temple unclothed in the boom
And, spontaneous, assumes to sing the soul,
As devotional hymn, in the shape of gold,
On the tempered strings of his,
The greatest lyre![9]

What I take the poet suggests to us, here, as does the violinist, is that we, cognitive scientists, work toward a definition of imagination and images allowing the scientific study of imagination not only to break with representationalism and internalism but also to revive, breaking with the mechanistic backdrop of representationalism and with the intellectualist backdrop of internalism, a theory of imagination and images, and, more largely, of cognition, as fantastic and magical.

Conclusion

Re-explaining Imagination

1. From a Methodological Point of View

1.1. A Starting Difficulty

Let me now go back from poetry to cognitive science. Let me say what studying and explaining imagination and images, according to what I said in Section 1, could concretely consist of. As I emphasized in Chapter 7, one could object that my approach to imagination leads to reductionism and eliminativism. Imagination would refer to cognition defined in embodied, enactive, and ecological terms, nothing less, nothing more. Worst, my book would provide interesting philosophical and poetical insights about imaginative aspects of cognition defined in terms of concrete and practical engagement with the world and in processes of making, producing, and so on. But it would provide no concrete explanation of imagination as we use to talk about it in contemporary cognitive science. I already explained why my approach is not reductionist. In the perspective I defend, the stake is not to eliminate imagination, to reduce it to specific aspects of cognition. Rather, it is (1) to make of imagination the condition and essence of cognitive experience and enablement and (2) to explain the diversity of imaginative practices and capacities in a way alternative to contemporary conceptions of imagination.

Let me clarify this. The question is to know how to explain imagination and how to study it concretely. Fine, but let's start over, from the beginning.

What is imagination?

As I emphasized earlier, many contemporary cognitive researchers admit that imagination is an ill-defined notion (Moran 1994, p. 106; Strawson, 1970, p. 31). Stevenson (2003), for example, counts twelve distinct conceptions of imagination at work in philosophy:

(1) The ability to think of something not presently perceived, but spatially and temporally real

(2) The ability to think of whatever one acknowledges as possible in the spatial and temporal world
(3) The ability to think of something that the subject believes to be real, but which is not
(4) The ability to think of things that one conceives of as fictional
(5) The ability to entertain mental images
(6) The ability to think of anything at all
(7) The nonrational operations of the mind, that is, those explicable in terms of causes rather than reasons
(8) The ability to form perceptual beliefs about public objects in space and time
(9) The ability to sensuously appreciate works of art or objects of natural beauty without classifying them under concepts or thinking of them as useful
(10) The ability to create works of art that encourage such sensuous appreciation
(11) The ability to appreciate things that are expressive or revelatory of the meaning of human life
(12) The ability to create works of art that express something deep about the meaning of life

Given this diversity of definitions, and in order to rationalize their study of imagination, cognitive researchers draw distinctions, between different kinds of imagination. To mention few of these distinctions, some of which I already introduced in Chapter 1:

- Propositional/sensory imagination (Stock 2017)
- Recreative imagining/creative imagining (Currie & Ravenscroft 2002)
- Sympathetic imagining/perceptual imagining (Nagel 1974)
- Enactive imagining/suppositional imagining (Goldman 2006a)
- Constructive imagining/attitudinal and imagistic imagining (Van Leeuwen 2013)
- Imagining from the inside/from the outside (Peacocke 1985, Shoemaker 1968)
- Imagining proper/supposing and conceiving (Balcerak Jackson 2016; Chalmers 2002)
- Experiential imagination (Dokic and Arcangeli 2015)

To each type of imagination corresponds a plethoric literature and, at the same time, a lack of consensus about what imagination concretely is. According to Peter Langland Hassan, again:

> in the case of imagination, "there doesn't even seem to be consensus about what the phenomenon under discussion is", much less agreement concerning its deepest nature. In trying to characterize "the phenomenon" of imagination, comparisons are made between imagination and states like perception and belief; but it's emphasized that imagination remains quite distinct from those states. Attempts to specify the precise ways in which it is distinct—and to thereby distinguish what it is we aim to study—threaten to leave us knee-deep in theory, before we've clearly identified what the theory is supposed to be theory *of*. (Langland-Hassan 2020, p. 3; quoting Kind and Kung 2016, p. 3)

Then, asking what a concrete study of imagination should consist in, presupposes we find a way to deal with this definitional difficulty. Does my approach of imagination help in any way? I take it that it does, by suggesting a specific methodological choice. In order to be specific about this methodological choice, let me compare my approach to Peter Langland-Hassan's approach. Langland-Hassan proposes we find a way to understand the diverse set of facts and capacities generally related to imagination, "by showing them to be instances of a single core pattern" (Langland-Hassan 2020, referring to Klitcher 1981, p. 530). The question is to know how the explanation of imagination can be unified within a broader framework for understanding the very nature of the mind. The idea is to break the complex and protean phenomenon of imagination into "simpler, more general parts" (Langland-Hassan 2020, p. 9). We need to find, as Langland-Hassan puts it, "unifying primitive" terms allowing us to explain imaginative acts and capacities as a phenomenon of mind. As for me, I totally agree with this way of thinking. I elaborated those primitive terms in the language of Simondon, pragmatism, material anthropology, and 5E approaches. In a way close to Peter Langland-Hassan's, I explained imagination according to an explanative framework allowing to understand how imagination and mind essentially relate to each other. But contrary to Langland-Hassan, I refused to adopt the classical internalist and representationalist framework. I gave reasons to abandon this framework and worked on elaborating and illustrating an alternative one.

Clearly, I endorse an eliminativism about imagination as a faculty, but a more pragmatist and analytical approach to imagination*s* understood as technically enabled capacities and dispositions. I deny the existence of a specific imaginative faculty, separated from other cognitive faculties. Imagination is not a faculty substantially different from other cognitive and conative faculties like perception, judgment, desire, hope, and so on. As I suggested some lines earlier, imagination is a doing and a volitional act. Dreaming is, in a way, hoping for and, at the same, realizing or failing to realize something. To separate imagination from other cognitive and conative activities makes no sense once we refuse to treat imagination as an independent faculty. And this is Langland-Hassan's core idea. However, instead of explaining different imaginative capacities and activities (dream, daydream, pretending, considering different plans for action, making up a story, enjoying fiction, picturing oneself in a situation, imagining how a tune goes, and so on) in terms of other and "well understood" (Langland-Hassan says) but intrinsically nonimaginative folk psychological states and processes, as Langland-Hassan proposes (2020 p. 10), I propose we explain these capacities and activities in terms of technical constitution, as cultural individualizations of imagination, as I defined it in Chapter 7 and throughout this whole book. Again, as I emphasized with Dewey in the introduction, against any idealistic temptation to talk about imagination in the absolute and as nonsituated, we need to acknowledge the multiplicity of imaginative situations being opened up by specific technical devices.

Concretely, what are the benefits of this methodology?

1. First, it gives a methodological framework for positively studying how, and explaining why, the differences mentioned above between a variety of imaginative activities and between a variety of types of imaginations are enacted. The idea is to study, whether in experimental or nonexperimental contexts, and from a developmental as well as from an evolutionary perspective, the emergence of imaginative capacities and activities. This would avoid studying imagination against the backdrop of unjustified theoretical and experimental frameworks and distinctions, unjustified in that they postulate the existence of categories of imaginative phenomena without being able to account for the origin and conditions of development of these phenomena. In short, approaching imagination this way helps one carefully build the notion one intends to study, whether in experimental contexts or, in a more ethnographical manner, in concrete practical contexts. Instead of staying in the lurch of arbitrary descriptive distinctions, we might be able to say what makes these differences, and why and how differences are enacted. By doing so, in a Simondonian verve, we would be able to think of each imaginative capacity and activity in the light of an explanation of its genesis—that is, in a more complete way.
2. Second, this approach to imagination allows to study and explain the relation between imagination and cognition on the one hand and technics

on the other. As I emphasized from the beginning, this is precisely what a definition of imagination in terms of representational mental states avoid thinking.

3. In the wake of (1) and (2), this approach suggests we might be able to put order in the study of imagination with regard to another science, that of technology. In the perspective I defended, indeed, the psychology and scientific study of imagination becomes inseparable from technology. It is in this sense that Bernard Stiegler (see Chapter 5) talked of "organology." Here again, the psychology of imagination becomes a kind of reflexive psychology or archeology of the mind, sort of a backward reflection on past imaginative and inventive processes through the study of given technical traces. It might, then, be possible to rebuild the definitional and explanative frameworks used in the cognitive science of imagination, in the light of such a technological approach to imaginative phenomena.
4. Finally, from a historical point of view, taking seriously (1) the imaginative nature of cognition and (2) the material and technical constitution of imaginations amounts to acknowledging the historical and technical relativity and accidentality of what we usually call "imagination." This allows to get rid of the referentialist bias I denounced in the introduction, according to which folk psychological concepts (historically and epistemologically situated descriptive constructs) refer to things in the world. In this sense, instead of taking, as Langland-Hassan did, folk psychological concepts for granted, and making an arbitrary distinction between purposedly well-understood, nonimaginative, and ill-understood imaginative folk psychological concepts, one should be able to face the diversity of imaginative acts and capacities without falling, as Langland-Hassan, into the easy trap of a reductive, still very classical, approach.

1.2. Beyond the Philosophy of Imagination

But here lies an ambiguity to be removed: Does this mean that imagination is reducible to technics and that studying imagination amounts to merely studying technical objects, milieus, and concretization processes? No, for sure, reducing psychology to technology is something we want to avoid. But again, and reversely, the philosophical and anthropological approach I offer, makes me believe that the classical assumptions about the intracranial ontological boundaries of human cognition and imagination should be resisted. Then, where to look? As put by Malafouris, "where is human cognition, and where should we look for it?"

(Malafouris 2013, p. 37). Once we reject, as I did, neuro-centrism, and refuse to locate cognition and, then, imagination in intracranial boundaries, where and how do we study imagination concretely? What should the study of imagination and our analyses focus on?

Here I would like to emphasize with Malafouris that studying cognition and imagination is studying *interactions*. Let us study imagination in Amandine Beyer's classes for examples. Whether we understand imagination as a particular case of cognition or as the general and conditional principle of cognition—or, as I suggested in Section 1, as the very essence and constant basis of cognition—the study of imagination focuses on interactions. I theorized this notion of interactions in the terms of Dewey's conception of experience as transactional, of Simondon's conception of transductive individuation, of Ingold's theory of transducers and correspondence, and in the language of 5E approaches. In the light of Malafouris' MET, I emphasized that thinking of cognition and imagination in terms of interactions does not amount to reducing the study of imagination to purely historical considerations. Organology and technology, for example, are, to appropriate Malafouris' words, primarily preoccupied with questions about the when and the where. Organologists are historians of musical evolution, not psychologists of musical imagination and creation. By contrast, what Malafouris' MET asks, which makes it a domain of cognitive science and psychology, and, as Gallagher emphasizes (Gallagher 2017; see Chapter 5), a starting point for any enactive and ecological study of imagination includes questions about the what, the why, and the how:

> What is symbolic thinking? Why and how did symbolism emerge? What forms of signification count as symbolic *meta*-representational thinking? Knowing when and where things are happening in cognitive evolution is important and interesting but does not explain much in itself. What we need is an integrative comparative perspective able to identify the different ingredients of cognitive change (evolutionary and developmental) and the causal mechanisms that underlie them in different contexts of human cognitive becoming. (Malafouris 2013, p. 39)

First remark then, a study of imagination as interaction would consist in analyzing and explaining the mechanisms underlying interactional phenomena, as they shape the imaginative and creative mind. As for me, I explained these mechanisms in terms of schematism. I do not pretend this is the only, or the better way. At least, it was a way for me to resituate the problem of imagination in the philosophical history of its formulation. I appropriated and combined theoretical and conceptual frameworks pertaining to the French philosophy of technics and to 5E approaches in order to take the philosophy of schematism a step further. This allowed me to explain in the language of cognitive science how imagination as interaction and the mechanisms of cognitive enablement underlying it ultimately work.

Fine, but again, what does it consist in, concretely, to study imagination and those mechanisms? And, first and foremost, why give so much importance, as I did in this book, to philosophy? Is it even necessary? In his review of Hutto and Myin's *Evolving enactivism* (2017), Evan Thompson remarks that philosophical arguments and criticisms of other theories do not make cognitive science. Instead, says he, we need to start from basic theoretical and empirical issues, those motivating the careful construction of positive theoretical frameworks and testable models. This being said, one cannot ignore—and Thompson, who contributed as a philosopher to the foundation of enaction on Merleau-Pontian grounds, does not ignore it—those theoretical frameworks, as they structure the ways we develop empirical and experimental designs, as well as the ways we interpret empirical data or the results of our experiments, take place in the course of a theoretical and conceptual history that is in itself, at least partly, philosophical. As Gaston Bachelard emphasized (1934)—as well as Karl Popper (1934) in the wake of David Hume's criticism of inductivism—scientific facts are built in the light of theoretical problematics and frameworks. Scientific discovery does not proceed through simple observation. Science goes against the illusion of immediate knowledge. Beyond the classical and obsolete opposition between empiricism and rationalism, scientific activity consists of an "applied" or "dialectical" rationalism. "No empty rationality, no disjointed empiricism" (Bachelard 1951, 1953). Science requires that the scientist transforms her rationality, theories, and concepts, on the basis of reality as it is experienced or experimented. Reversely, scientific experiments, as instrumented, technically constituted situations, study phenomena to the only extent that these phenomena are built theoretically, as objects of knowledge, in the context not only of historically situated theoretical and conceptual but also of technical frameworks (see Bachelard's notion of phenomenotechnics). In schematic dialectical terms, one could say that a proper scientific study of imagination as interaction would consist in (1) building theoretical and explanative frameworks and conceptual tools in the light of which (2) designing concrete experimental situations in the flesh of which (3) cognitive scientists would transform their explanations (theoretical frameworks and conceptual tools) of imagination. Along this way, philosophy, understood as an exegetic activity of theoretical and conceptual problematization and clarification, far from leading us to a useless scholasticism, has its place in science, in our case in a cognitive science of imagination.

In this sense, if a cognitive philosophy is even to have a function in cognitive science, it is to serve as an epistemological and ontological clarification in the light of which conducting experimental research. In this book, I systematically associated philosophical exegesis, conceptual clarification, and ethnographic description. I made my theoretical and conceptual framework evolving in the flesh of ethnographical descriptions, among others, cello bowing and orchestral rehearsal. But I did not engage in experimental design and quantitative research. My theoretical edifice, as says Malafouris about his MET (2013 p. 51), should not be confused with a positivistic scientific theory, and has no pretention to serve as a predictive theory. Not that my approach to imagination is not amenable to rigorous empirical testing research. Simply, this would be the next step of a research path, a path that needs to be philosophically grounded and enlightened. In this sense, the enactive and ecological approach to imagination I offered in this book is more a preliminary research program, an explanatory framework to start with, alternative to the classical internalist and representationalist framework I rejected in order to study the relation between imagination and technics.

1.3. The Developmental Perspective

Interactions basically refer to how the brain, the body and the material and technical world relate to each other in the human cognitive becoming and shaping. In the perspective I defend, studying imagination as interaction would amount to studying the co-evolution of imaginative acts and capacities on the one hand and of brains, bodies, and technical (instrumental, technological, digital, institutional, and so on) environments on the other. The idea would be to study this co-evolution on the scale of the individual and of the collective, on short-term and long-term periods of time, at the scale of human evolution as well. Theoretical and methodological models giving due importance to external factors and to situated experience in the explanation of individual's psychological development would be of special interest. As suggested by Malafouris (2013, p. 40), probabilistic epigenesis offers promising perspectives. Probabilistic epigenesis is an integrative, bidirectional, and nonhierarchical (Cohen 2009) explanative model of development. This model assumes that neural structures derive from the organism intrinsic activity and developmental dynamisms, as well as from extrinsic interactions (Gottlieb 2002, 2003, 2007; Cohen 2009). Development involves bidirectional interactions between equally determining biological, psychological, and social levels. Combining the theoretical and methodological framework of probabilistic epigenesis with an enactive and ecological approach to imagination as technically constituted would amount to giving importance not only to biological and social but also to technical factors in the development (enablement, enactment) of imaginative acts and capacities.

From such an integrative developmental perspective, indeed, the study of individuals' dispositions to perceive, act, and think, as they structure individuals' psychological and emotional engagement with the world and with others, and as they include and develop in the form of imaginative capacities like anticipation, pretense, empathy, hypothetical reasoning, and so on, would include the empirical and experimental study of imagination understood in terms of human-technics interaction. In a Piagetian verve, one would study the development of the imaginative and epistemic subject and explain it in terms of interactions and engagements in socio-technical environments. In a more Wallonian perspective, one would study the development of emotionality and of social bonding, as constituted not only, but still essentially, through technically constituted imaginative practices. The explanation of children and adolescents' emotional, social, and cognitive development or disorders (Cohen 2009), for example, would include the study of children and adolescents' engagements in specific technical contexts. More specifically, the study would focus on the role of specific technical mediations and environments in the enactment of specific kinds of imaginative capacities, these capacities being involved in children and adolescents' social or cognitive life and development. Examples of such approaches already exist. To take but one example: the role of digital technologies in the development of attention (e.g., Hopkins, Brookes, and Green 2013; Rideout and Hamel 2006; Mills 2014; Meshi, Tamir, and Heekeren 2015; Loh and Kanai 2016; Uncapher et al. 2017; Vedechkina and Borgonovi 2021). Beyond alarmist or optimistic views, the stake would be to study and explain, in the light of, and in discussion with, the enactive and ecological framework, the causal correlations between brains, individuals' as well as collectives' behaviors and dispositions, imaginative capacities, and technical environments. In an even more normative perspective, the challenge would also be to discern the possible ways in which the actual nature of the relationship between these dimensions of human individuation could be oriented, according to a clear understanding of the kind of capacities and dispositions one wants to promote and develop (education studies).

In this sense, studying imagination "as interaction" has nothing to do, of course, with interaction design, as it is associated, in Kaptelinin and Nardi (2006) for example, with activity theory. Interaction design refers to the understanding of human engagement with digital technologies and the effort to use that knowledge "to design more useful and pleasing artifacts" (Kaptelinin and Nardi (2006 p. 5). My approach to the enactive relation between imagination and technics could certainly benefit those who work in the fields of human-computer interaction, computer-supported interactive and collaborative work and learning, digital design, and cognitive ergonomics. But instead of adopting a user-centered view, as if users were already constituted as cognitive entities, and as if the only challenge was to make user interfaces more efficient, pleasing, the stake would be to think of the kind of cognitive entities one wants to enact through these technologies, on ethical, moral, social, artistical, political, and intellectual levels, for example. Again, in the perspective I adopt, human intelligence is not situated and interactional in the mere "inter-active" sense of the term. Beyond mere interoperability—the idea that human operations and technical operations interact and constitute coupled operative systems—there is, again, enablement, enaction, and discovery of signification (Simondon). Intelligence is situated in a more "intra-active" than interactive sense (Malafouris 2013, p. 39). It shapes and constitutes human intelligence. And if interactional technologies like user interfaces shape the ways we relate to ourselves and to each other (see on this Robertson 2006), and participate in the enactment of dispositions and of perceptive worlds, then, beyond pleasing and useful interfaces, who do we want to become, and what worlds do we want to live in? How do we imagine ourselves and our worlds?

From a historical or archeological perspective, this study would focus on the relation between technical achievements and imaginaries (historically situated sets of imaginings), as they combine in concrete practical (political, religious, juridic, educational, scientific, artistical, and so on) contexts and dispositions to perceive, act and think. The stake would be to describe and explain long-term change and, more particularly, historical correlations between technical environments and imaginaries as they structure political, religious, juridic, educational, scientific, artistical, and so on practices. As regards scientific activity, this would align with the French school of historical epistemology, among others, with the works of Bachelard (1938) and Canguilhem (1977).

2. From an Epistemological Point of View

2.1. A New Taxonomy? Making Distinctions Instead of Confusions

I proposed we make a distinction between sensory imagery and imagination. I suggested that, when talking about sensory imagery in terms of imagination, we make a confusion between something that is not imaginative, namely sensory imagery on the one side and images and imagination properly speaking on the other side. Drawing on this, I said that mental images, as representational entities in the mind, do not exist. I also gave much attention to the notion of experiential imagination. I argued that imagination is experiential, in an absolutely essential way. Imagination is part and condition of experience. And I said very little about propositional imagination. In this subsection, I will (1) summarize quickly these key ideas, (2) explain why I had, by contrast with contemporary approaches of imagination, to put propositional imagination aside, and (3) say what my approach to imagination involves with respect to the common taxonomy distinguishing sensory, experiential, and propositional imagination. In doing so, in a conclusive subsection, I will define images precisely, and say why it was relevant to think from the beginning in terms of schematism.

Let me start, so as to make my view as clear as possible with Langland-Hassan's distinction between imagistic imagining (i-imagining) and attitude imagining (a-imagining). According to Langland-Hassan, i-imaginings have a sensory character, correspond to situations we usually describe in terms of "mentally visualizing" or "seeing an image in our mind's eye": "where, again, the 'image' in 'imagine' is emphasized" (Langland-Hassan 2020, p. 5). In this book, I offered elaborated reasons to think that an easy reduction of the notion of "image" to the notion of sensorial "imagery," as if language biases were enough to legitimate a concept, is misleading. Again, coming back to what I argued in Chapter 2 in the light of pragmatism and of Bennett and Hacker's Wittgensteinian approach, mental visualizations, as realized in the visual cortex in the form of isomorphic processes, are no images of anything. I will not dwell on this again. Instead, I shall say exactly what an image is, by difference with sensory imagery.

Worth noting, Langland-Hassan denies that sensory imagery counts as representational. I remember a personal conversation I had with him in his office at Cincinnati University, where he told me that in order for an image of someone looking exactly like my grandmother, to be an image of my grandmother, this image needs to count as a representation. And the only way, he told me, for this sensory image to count as a representation is to add propositional content to it. In this sense, in his view, i-imaginings merely refers to the "use of mental imagery in thought" (Langland-Hassan 2020, p. 5). If one wants to speak about imagination in terms of elaborated cognition about the possible, of fantastical pretended, of a faculty to mentally conceive, assume, suppose, devise, plot, compass, consider, ponder, mediate, and so on—in short, if one wants to talk about imagination as a representational faculty, one needs to postulate, beyond i-imaginings, a-imaginings. A-imaginings refer to the propositional or "cognitive" (in the restricted and classical sense of computo-symbolic) attitude of imagining, "a sui generis psychological mode of imagining" (Langland-Hassan 2020, p. 5). Thinking about how imaginings can have correctness conditions, Langland-Hassan assumes that imaginings represent thanks to their nonimagistic, propositional, content (Langland-Hassan 2015). The specific portion of an imagining contributed by a sensory image (static images and sequences of images) has in itself no representational function. Imaginings, in order to be judgments about the world ("this is my grandmother and not a deepfake"), need to combine with nonimagistic but language-like content.

As I emphasized earlier, the idea that the sensorial portion of cognitive experience needs, in order to count as representation, to be associated with something else, a concept for example, is not new. This, again, is the very problem of schematism, as it is formulated in Kant and as it takes root before Kant in a whole philosophical history that goes back to Aristotle, that preoccupies the whole scholastic tradition and, breaking with scholasticism, Descartes. The question is how to conceive that a sensory image has meaning, and corresponds to a concept that we can manipulate in logical, true, or false judgments about the world? Langland-Hassan endorses the propositionalist solution. As for me, as I did not cease to suggest throughout this whole reflection, I tend to think that solving this massive philosophical problem of representation by appealing to propositional content and mental representations, as defined in cognitive science, is nothing but a philosophical shortcut that lies in the easy presupposition of the very phenomenon we want to explain, namely representation (mereological fallacy; see Chapter 2). In the previous chapters, I worked on explaining the possibility and the nature of representational acts, whereby sensory imagery participates in meaningful experiences, in the terms of schematism, schematism defined as experience, in terms of embodied, enactive, and ecological engagement. In Chapter 6 I emphasized that correctness conditions depend on the organism's engagement in practical and technical contexts. In short, sensory imagery has nothing to do with images properly speaking. And to be an image of something, a process needs to be embodied and situated, technically constituted, temporally extended.

To conclude on this, in order to think of the exact nature and cognitive function of sensorial imagery and of propositional imagination (capacity to imagine in the form of propositions), I thematized the notion of experiential imagination. I said that imagination is essentially experiential, has to do with experience defined in transactional, relational, embodied, enactive, and ecological terms. I made of experience defined as a process of technically mediated worldly engagement and cognitive enablement (schematism), the conditional dynamism through which our abilities to imagine in the form of visual, auditory, or propositional imaginings, for example, develop. If one can imagine in the form of propositions, it is because one engages in technically constituted writing and reading practices, not in virtue of any propositional essence of the mind. If one can imagine in a musical language, it is because one engages in musical practices, not in virtue of any musical essence of the mind. If one can imagine without words and privately visualize fictional things, it is because one engages in practical, technically mediated activities of visual image production, not in virtue of a visual essence of the mind. Aristotle made of vision the primary and most delightful sense of humans (*Metaphysics*, Alpha, 980a). But our ability to imagine visually is not pregiven. It is constituted through technical engagement. To have private and meaningful visual imaginings depends on our constant engagement with technical devices of visual perception (see, for example, Azéma and Rivère 2012, on the Paleolithic and technical constitution of our ability to represent and imagine movement; see also White (1973) on how visual imagination in the nineteenth century developed through the concrete invention and use of optical theaters).

In this sense, I think that the taxonomy distinguishing sensory, experiential, and propositional imaginations, as it is usually presented (Kind 2016), is arbitrary and obsolete. I suggest that we make of imagination as experiential the basis of our embodied, enactive, and ecological inquiries on imagination. And I suggest that the other types of imagination be conceived in relation with the specific domains of technicity in which they are enacted and in which they develop. No propositional imperialism, that is, not necessity to make, as classical cognitive science and Langland-Hassan, of propositional imagination the only alternative to sensorial imagery and the essence of all our imaginative capacities. This makes no sense. No empire in an empire. The only empire is that of life and imagination, of imagination as life and as a technically constituted engagement with the world.

2.2. What and Where Are Images? How to Study Them?

Now, what are images? And where are they, if not in the brain? Where to look for them and how to study them? To answer this question, one needs to say what characterizes something as an image. To summarize the key idea I developed on this matter, an image is an image to the extent that it is a way for me to relate in a representational manner to something. In the language and ontology I elaborated here, then, images are relational or ecological; they refer to coupling structures between the individual and the world, whereby meaning is enacted and instantiated in representational acts. Images, in this perspective, refer to relational schemes. They are not on the side of interiority (a representational brain or disembodied mind), nor on the side of pure exteriority (mere external representations), as if a photography, for example, was in itself an image of something. Rather, images are on both sides at the same time, they exist as meaningful and relational and compositional (Bruno) engagement processes. Images are ecological in nature.

This way of talking goes against the traditional conception of images as pertaining to the private interiority of an imaginative subject. As a matter of fact, with Simondon, in the prolongation of Bergson and in the sense of a radical rejection of the phenomenological approach, I insisted on the idea that the image does not belong to the only register of the subjective interiority. The rejection of the phenomenological prism leads Simondon, in the prolongation of Bergson, to adopt from the very first lines of his lessons on imagination and invention, a realistic and radically anti-subjectivist posture. Simondon positions himself against the attitude too common in philosophy that consists in defining the imagination as a faculty, as a subjective power, in favor of what one could call a "primitive externalism of the image." The error is indeed, according to him, to "attach the images to the subject that produces them", and "to exclude the hypothesis of a primitive exteriority of the images with regard to the subject." Explicitly against Sartre, against his definition of the imagination in terms of imagining consciousness, Simondon underlines the independence and the primary externality of images. No subjective power, whether representational or not, to produce mental images. The image proceeds, rather, from a relation directed by the exteriority and by the own forces, self-determining and self-constituting of the images:

> But why exclude as illusory the characters by which an image resists free will, refuses to let itself be directed by the will of the subject, and presents itself according to its own forces, inhabiting the consciousness like an intruder who

> comes to disturb the order of a house where he is not invited? (Simondon 1965–66, p. 7)

In Simondon's perspective, images have their own power, irreducible to subjective interiority:

> The dream, with the dreamlike figures that animate it, is not only what we would call a subjective event; it manifests a power, an intention, a reality that does not have its source in the subject but that, on the contrary, comes to him and seeks him out. The image that invades the subject is an apparition. (Simondon 1965–66, p. 8)

In this sense, emphasizing, as I did in this book, the ecological nature of cognition in general and of imagination and images in particular is at the same time acknowledging something very simple: studying brain imagery in no way is studying images and imagination. Images and imagination carry, as ontic and dynamic parts of themselves, the constant body-world relation.

Instead of situating images in pure exteriority, which Sartre denounces as a metaphysical shortcut (see Chapter 6) and which leads to reduce images to material images, I defined images as relational phenomena, as relational schemes. They are in the constant relation that constitutes the individual-world system, the enactive and ecological coupling between the organism and the environment. And as I suggested, the study of images defined this way requires focus on this interactional, intra-active, or relational nature of images. Following Bruno's insights about the compositional nature of images and meaning, images are not just compositions of simulacra out there in the world. Rather, they are mirrors, that is, images to the only extent that, as goes the metaphor of mirror in Bruno or in Nicolas of Cusa as well (see Chapter 2, footnote), we reflect in it as meaningful minds. Put differently, images are compositional sets of simulacra whereby meaning, as a relation between the *mens* and the world, is enacted. As such, images cannot be reduced to their material part. And their psychological part cannot be thought independently from its constitution through the externality of material images. Again, studying imagination and images demands we work in the sense of an anthropology of the fantastic, where psychology and technology (science of the history of technological coupling devices) combine to make us believe in magic.

Notes

Chapter 2

1 In this sense, epistemology has a vocation to highlight what Gaston Bachelard, French philosopher of sciences, calls "epistemological obstacles," that is, the confusions and metaphorical traps that intrinsically constitute the very acts thought and scientific knowledge (Bachelard 1938). For a comparison between Bachelard's and Dewey's epistemologies, see Fabre, M. (2005): https://www.cairn.info/revue-les-sciences-de-l-education-pour-l-ere-nouvelle-2005-3-page-53.htm).

2 https://www.psychologytoday.com/us/blog/waking-dreaming-being/201504/the-brain-is-wider-the-sky.

3 Here, a little detour through Nicolas of Cusa's mirror metaphor, a metaphor I will come back to in Chapter 7, could be relevant. Emily Dickinson's poem somehow points to what philosophers of the Renaissance defined as the unknowable and the ineffable. Scholasticism, of neoplatonic as well as peripatetic inspiration, used to oppose the infinite nature of the spiritual world (mind, soul, Plato's soul of the world, God) to the finite nature of the material world (sublunar world, bodies). The question was to know how to reunite these two worlds. For scholastics of the thirteenth and fourteenth centuries, for example (Saint Thomas Aquinas, John Duns Scotus, William of Ockham), the stake was to explain how both worlds relate to each other: ontologically through a hylomorphic process of form generations on the one hand and epistemically through a philosophical and conceptual process of unveiling and revealing on the other hand. God was to be known, revealed through philosophy, in the light of Plato and Aristotle's conceptual and metaphysical frameworks.

In the fifteenth century, humanism and modern philosophers started challenging scholasticism. Among others, Nicolas of Cusa breaks with the scholastic tradition, assuming that there is no way for the human mind to know God. There is no concept that could help us know and understand God:

> Thy face, so long as he forms any concept thereof, is far from Thy face. For all concept of a face falls short, Lord, of Thy face . . .. In all faces, is seen the Face of faces, veiled, and in a riddle; howbeit unveiled it is not seen, until above all faces a man enters into a certain secret and mystic silence where there is no knowledge or concept of a face. This mist, cloud, darkness or ignorance into which that seeks Thy face enters when he goes beyond all knowledge or

> concept, is the state below which Thy face cannot be found except veiled; but that very darkness reveals Thy face to be there, beyond all veils. (Nicolas of Cusa 1453/2007, pp. 26, 27)

God is unknowable. We cannot conceive Him. As says Nicolas of Cusa, God is hidden behind "the wall of Paradise" (1453/2007, p. 49). Instead of conceiving God, however, we can relate to Him through a specific feeling, that of the ineffable. Something ineffable is something that, given its nature and intensity, cannot be conceived nor said. It is something that can neither be explained nor apprehended by the only reason. The ineffable is something we experience in feeling. And according to Nicolas of Cusa, this feeling of God, feeling of the ineffable, combines the feeling of our ignorance and the feeling of God's incommensurability.

Ignorance for Nicolas of Cusa is wise (*De Docta ignorantia*, 1440) in the sense that acknowledging the unknowable nature of God is the only way to relate to Him. God exceeds our understanding in ways (riddles, veils, mist) we even cannot understand. God is incomprehensibly incomprehensible. He doubly exceeds our understanding. Relating to God, in this sense, is a philosophical and Socratic as well as an ecstatic act par excellence.

Nicolas of Cusa addresses, then, a paradoxical question, which is to know how to know what is unknowable. And he answers as follows: relating to God consists not in conceiving Him, but in feeling a contrast between two orders of magnitude, between two different and incommensurable scales: the Face of God and faces, the infinite and the finite, the absolute knowledge and our ignorance. Brain and God, Brain and Sky, Brain and Ocean: each of Emily Dickinson's images points to this incommensurability. As I said, Dickinson's poem suggests that, as minds, we extend to the scale of God, of the sky, and of the sea. The brain is absorbed as much as it absorbs. But if we take inspiration from Nicolas of Cusa's mirror metaphor, by means of which he explains the feeling of the ineffable, then, the brain *absorbs to the only extent* that it is absorbed. The brain contains *to the only extent* that it is contained. In his *Vision of God* (1453), indeed, Nicolas of Cusa writes:

> Now I behold as in a mirror, in an icon, in a riddle, life eternal, for that is naught other than that blessed regard. (Nicolas of Cusa 1453/2007, p. 17)

God reflects in creatures as faces reflect in mirrors (see Viau 2009). Finite entities are mirrors of the infinite Entity. Faces reflect, as mirrors, the Face of God. Here, as I will emphasize in Chapter 7 with another thinker of the Renaissance, Giordano Bruno, the role and nature of images is in question. Images allow us, in Nicolas of Cusa's perspective, to know what is unknown. They allow us to know God. They allow incommensurable scales to relate to each other. But, at the same time, as it is the case in the Platonic metaphysic, as well as in Nicolas of Cusa's, images, as reflections of the being, never exhibit the entire nature and reality of what they reflect. What is reflected exceeds, ontologically and epistemically speaking, its reflection. And images, as reflections, depend, in order to exist as such reflections, on that which they reflect, that which they mirror. This fundamentally differs, as I will show in Chapter 7, from Giordano Bruno's conception of images as mirrors. However, coming back to the brain-mind relation, in a Cusian perspective, the brain *is* not the mind. Instead, the brain *reflects* the mind. It is, metaphorically speaking, a mind, to the only extent that it reflects the mind of which it is part, on which it depends, in order to exist.

Chapter 3

1 In 1952, Simondon intended to devote his PhD dissertation to the notion of individuality and his complementary dissertation to the psycho-physiological relation; his 1958 PhD thesis masterfully combines these two questions.
2 Finally published in *Chiasmi International*, vol 7.
3 See Nathalie Simondon, http://gilbert.simondon.fr/content/biography.

Chapter 4

1 In the pictorial sense of what illuminates from the inside.
2 As concerns the way Whitehead translate, with his concept "concrescence," bergsonian ideas of "durée" and "progrès," see Didier Debaise, "Devenirs et individuations. L'hommage de Whitehead à Bergson. https://journals.openedition.org/noesis/1637.
3 Pallasmaa, J. 2009. *The Thinking Hand: Existential and Embodied Wisdom in Architecture*. Chichester: Wiley, p. 111.
4 Cited by Ingold 2013a, p. 128. Pallasmaa, J. 2009. *The Thinking Hand: Existential and Embodied Wisdom in Architecture*. Chichester: Wiley, p. 17.

Chapter 6

1 Here is a little video I made on a conversation with Tormod Dalen about the embodied and instrumental nature of musical imagination : https://www.youtube.com/watch?v=Uw9a-Q3FT7M.

Chapter 7

1 *Papillon* (2020). This is my first film, it was selected and projected in two festivals in Paris (*Athmosphères* and *Urban Film Festival*), and it was awarded with a special mention in the festival *Athmosphères*. https://www.youtube.com/watch?v=CN8Jx5h7Qpk.

2 Excerpt from the jury's speech at the award ceremony, festival *Athmosphères*.

3 In this sense, the question is not to know how individual may become more imaginative and more fanciful. Instead, it is to know how, in which technical and institutional contexts, individuals learn to separate their actions, perceptions, and thought, from their imagination and fantasy. It is to explain, from a developmental and anthropological perspective, the distinction between imagination on the one hand and perception, action, and thought on the other—in one word, cognition.

4 https://www.youtube.com/watch?v=k1lLQYYmBfs

5 « Notre appartenance au monde des images est plus forte, plus constitutive de notre être que notre appartenance au monde des idées ».

6 « Oculus qui ita videt alia ut et se videat . . . quique est ommia idem ».

7 « Noster oculus se ipsum cerneret ».

8 « Non dicuntur imagines ex ratione rerum significatarum, sed potius ex ipsarum significantium conditione »

9 Je compose en esprit, sous les myrtes, Orphée
L'Admirable ! . . . Le feu, des cirques purs descend ;
Il change le mont chauve en auguste trophée
D'où s'exhale d'un dieu l'acte retentissant.
Si le dieu chante, il rompt le site tout-puissant ;
Le soleil voit l'horreur du mouvement des pierres ;
Une plainte inouïe appelle éblouissants
Les hauts murs d'or harmonieux d'un sanctuaire.
Il chante, assis au bord du ciel splendide, Orphée !
Le roc marche, et trébuche ; et chaque pierre fée
Se sent un poids nouveau qui vers l'azur délire !
D'un Temple à demi nu le soir baigne l'essor,
Et soi-même il s'assemble et s'ordonne dans l'or
À l'âme immense du grand hymne sur la lyre !

Bibliography

Ansaldi, S. (2013). *L'Imagination fantastique. Images, ombres et miroirs à la Renaissance* (The Fantastic Imagination. Images, Shadows and Mirrors at the Renaissance). Paris: Les Belles Lettres.

Arandia, I. R. and Di Paolo, E. A. (2022). "Symptom Perception, Placebo Effects, and the Bayesian Brain." *PAIN*, 163, e604.

Assagioli, R. (1965). *Psychosynthesis: A Manual of Principles and Techniques*. New York and Beunos Aires: Hobbs, Dorman & Co., Inc.

Aydin, C. (2015). "The Artifactual Mind: Overcoming the 'Inside–Outside' Dualism in the Extended Mind Thesis and Recognizing the Technological Dimension of Cognition." *Phenomenology and the Cognitive Sciences*, 14(1), 73–94.

Azéma, M. and Rivère, F. (2012). "Animation in Paleolithic Art: A Pre-echo of Cinema." *Antiquity*, 86(332), 316–24.

Bachelard, G. (1934/2013). Le Nouvel esprit scientifique (The New Scientific Mind). Paris: Presses Universitaires de France.

Bachelard, G. (1938). *La Formation de l'esprit scientifique. Contribution à une psychanalyse de la connaissance objective*. Paris: Vrin.

Bachelard, G. (1941). *L'Eau et les rêves : Essai sur l'imagination de la matière* (Water and Dreams: Essay on the Imagination of Matter). Paris: José Corti.

Bachelard, G. (1943). *L'Air et les Songes : Essai sur l'imagination du mouvement* (Air and Dreams: Essay on the Imagination of Movement). Paris: José Corti.

Bachelard, G. (1951). *L'Activité rationaliste de la physique contemporaine* (The Rationalist Activity in Contemporary Physics). Paris: Presses Universitaires de France.

Bachelard, G. (1953). *Le Matérialisme rationnel* (The Rational Materialism). Paris: Presses Universitaires de France.

Bachelard, G. (1988). *Fragments d'une poétique du feu* (Fragments of a Poetic of Fire). Paris: PUF.

Baggs, E. and Chemero, T. (2018). "Radical Rmbodiment in Two Directions." *Gestalt Phenomenology and Embodied Cognitive Science*, 198(3), 2175–90.

Balcerak Jackson, M. (2016). "Justification by Imagination." In F. Macpherson and F. Dorsch (eds.), *Perceptual Imagination and Perceptual Memory*, 209–26. Oxford: Oxford University Press.

Barbaras, R. (2010). "Life and Exteriority: The Problem of Metabolism." In J. Stewart, O. Gapenne, and E. A. Di Paolo (eds.), *Enaction: Toward a New Paradigm for Cognitive Science*. Cambridge: MIT Press, 89-122 .

Bellis, D. (2020). “Vision, image et imagination chez Descartes et Gassendi” (Vision, Image and Imagination in Descartes and Gassendi). Les Cahiers philosophiques de Strasbourg, 165–92. https://doi.org/10.4000/cps.4502.

Bennett, M. R. and Hacker, P. M. S. (2006). *Philosophical Foundations of Neuroscience*. Maiden, MA: Blackwell Publishing.

Bennett, M. R., and Hacker, P. M. S. (2003/2014). *Philosophical Foundations of Neuroscience*, ed. P. M. S. Hacker. Hoboken, New Jersey: Wiley-Blackwell.

Bennett M. R., Dennett, D., Hacker, P. M. S., and Searle, J. (2007). *Neuroscience and Philosophy: Brain, Mind, and Language*. New York: Columbia University Press.

Bergson, H. (1889). *Essai sur les données immédiates de la conscience* (An Essay on the Immediate Data of Consciousness). Paris: Félix Alcan.

Bergson, H. (1896). *Matière et mémoire. Essai sur la relation du corps à l'esprit* (Matter and Memory. An Essay on the Relation of Body and Spirit). Paris: Félix Alcan.

Bergson, H. (1919/2009). *L'Énergie spirituelle* (The Spiritual Energy). Paris: Presses Universitaires de France.

Bernardi, N. F., Schories, A., Jabusch, H.-C., Colombo, B., and Altenmüller, E. (2013). “Mental Practice in Music Memorization: An Ecological-Empirical Study.” *Music Perception*, 30(3), 275–290.

Blakemore, C. (1977). *Mechanisms of the Mind*. Cambridge: Cambridge University Press.

Boden, M. A. (1990). *The Creative Mind: Myths and Mechanisms*. London: Weidenfeld & Nicholson.

Boden, M. A. (1994). “Précis of The Creative Mind: Myths and Mechanisms.” *Behavioural and Brain Sciences*, 17(3), 519–70.

Boden, M. A. (1995). “Creativity and Unpredictability.” *Stanford Education and Humanities Review*, 4(2), 123–39.

Boeda, E. (2021). *Le phénomène technique en préhistoire. Une réflexion épistémologique à partir et autour du Levallois* (The Technical Phenomenon in Prehistory. An Epistemological Reflection from and around the Levallois). Collection *Cahiers d'Anthropologie des Techniques*. Paris: L'Harmattan.

Bourgine, P. and Stewart, J. (2004). “Autopoiesis and Cognition.” *Artificial Life*, 10, 327–345.

Bower, G. H. (1970). “Imagery as a Relational Organizer in Associative Learning.” *Journal of Verbal Learning and Verbal Behavior*, 9(5), 529–33.

Bower, G. H. (1972). “Mental Imagery and Associative Learning.” In L. W. Gregg (ed.), *Cognition in Learning and Memory*, 51–88. New York: John Wiley & Sons.

Bower, M. and Gallagher, S. (2013). “Bodily Affects as Prenoetic Elements in Enactive Perception.” *Phenomenology and Mind*, 4(1), 78–93.

Brann, E. T. H. (1991). *The World of Imagination: Sum and Substance*. Savage, Md: Rowman & Littlefield.

Brooks, R. A. (1991). “Intelligence without Representation.” *Artificial Intelligence*, 47(1–3), 139–59.

Bruineberg, J. and Rietveld, E. (2014). "Self-Organization, Free Energy Minimization and Optimal Grip on a Field of Affordances." *Frontiers in Human Neuroscience*, 8, 599.

Bruno, G. (1591/2009). *De imaginum, signorum et idearum compositione* (On the Composition of Signs, Images and Ideas). New York: Willis, Locker & Owens.

Bryson, N. (2003). "A Walk for Walk's Sake." In C. de Zegher (ed.), *The Stage of Drawing: Gesture and Act*, 149–58. London: Tate Publishing; New York: The Drawing Center.

Bugelski, B. R. (1970). "Presentation Time and the Total-Time Hypothesis: A Methodological Amendment." *Journal of Experimental Psychology*, 84(3), 529–30.

Byrne, A. (2010). "Recollection, Perception, Imagination." *Philosophical Studies*, 148(1), 15–26.

Cabeza, R. and Nyberg, L. (2000). "Imaging Cognition II: An Empirical Review of 275 PET and fMRI Studies." *Journal of Cognitive Neuro science*, 12, 1–47.

Cahn, D. (2008). "The Effects of Varying Ratios of Physical and Mental Practice, and Task Difficulty on Performance of a Tonal Pattern." *Psychology of Music*, 36(2), 179–91.

Canguilhem, G. (1943/2009). *Le Normal et le pathologique* (The Normal and the Pathologic). Paris: Presses Univesitaires de France.

Canguilhem, G. (1952/1993). *La Connaissance de la vie* (The Knowledge of Life). Paris: Vrin.

Canguilhem, G. (1955). *La Formation du concept de réflexe aux xviie et xviiie siècles* (The Formation of the Concept of Reflex in the 17th and the 18th Century). Paris: Presses Universitaires de France.

Canguilhem, (1977) Idéologie et rationalité dans l'histoire des sciences de la vie, Éd. Vrin, coll. poche Bibliothèque des textes philosophiques, Paris, 2009.

Carruthers, P. (2002). "The Roots of Scientific Reasoning: Infancy, Modularity, and the Art of Tracking." In P. Carruthers, S. Stich, and M. Siegal (eds.), *The Cognitive Basis of Science*, 73–95. Cambridge: Cambridge University Press.

Carruthers, P. (2011). "Creative Action in Mind." *Philosophical Psychology*, 24(4), 437–61.

Casey, E. S. (1976/2000). *Imagining: A Phenomenological Study*. Bloomington: Indiana University Press.

Chalmers, D. (2002). "Does Conceivability Entail Possibility?" In T. S. Gendler and J. Hawthorne (eds.), *Conceivability and Possibility*, 145–200. Oxford: Oxford University Press.

Chemero, A. (2009). *Radical Embodied Cognitive Science*. Cambridge: The MIT Press.

Chittick, W. C. (1994). *Imaginal Worlds — Ibn al-ʿArabī and the Problem of Religious Diversity*. Albany: SUNY Press.

Churchland, P. (1986). *Toward a Unified Science of the Mind-Brain*. Cambridge, MA: MIT Press.

Clapin, H., ed. (2002). *Philosophy of Mental Representation*. Oxford: Clarendon Press.

Clark, A. (1997). *Being There: Putting Brain, Body, and World Together Again*. Cambridge, MA: MIT Press.

Clark, A., ed. (2008). *Supersizing the Mind: Embodiment, Action, and Cognitive Extension*. New York: Oxford University Press.

Clark, A. (2010). "Memento's Revenge: The Extended Mind, Extended." In R. Menary (ed.), *The Extended Mind*, 43–66. Cambridge, MA: MIT Press.

Clark, A. and Chalmers, D. (1998). "The Extended Mind." *Analysis*, 58(1), 7–19.

Clark, A. and Grush, R. (1999). "Towards a Cognitive Robotics." *Adaptive Behavior*, 7, 5–16.

Clark, A. and Toribio, J. (1994). "Doing Without Representing?" *Synthese*, 101, 401–31.

Clark, T., Williamon, A., and Aksentijevic, A. (2012). "Musical Imagery and Imagination: The Function, Measurement, and Application of Imagery Skills for Performance." In D. Hargreaves, D. Miell, and R. Macdonald (eds.), *Musical Imagination: Multidisciplinary Perspectives on Creativity, Performance, and Perception*, 351–65. Oxford: Oxford University Press.

Cohen, D. (2009). "Probabilistic Epigenesis: An Alternative Causal Model for Conduct Disorders in Children and Adolescents." *Neuroscience and Biobehavioral Reviews*, 34(1), 119–29.

Cohen, M. S., Kosslyn, S. M., Breiter, H. C., DiGirolamo, G. J., Thompson, W. L., Anderson, A. K., Brookheimer, S. Y., Rosen, B. R., and Belliveau, J. W. (1996). "Changes in Cortical Activity during Mental Rotation. A Mapping Study Using Functional MRI." *Brain*, 119(Pt 1), 89–100. https://doi.org/10.1093/brain/119.1.89.

Cox, A. (2016). *Music and Embodied Cognition: Listening, Moving, Feeling, and Thinking* (*Musical Meaning and Interpretation*). Bloomington: Indiana University Press.

Crick, F. (1995). *The Astonishing Hypothesis*. London: Touchstone Books.

Currie, G. and Ravenscroft, I. (2002). *Recreative Minds: Imagination in Philosophy and Psychology*. Oxford: Oxford University Press.

Dalen T. (2016). "Zum Spielen und zum Tantzen', a Kinaesthetic Exploration of the Bach Cello Suites through Studies in Baroque Choreography." *Journal for Artistic Research*, 11. http://doi.org/10.22501/jar.103350

Damasio, A. R. (2001). "Some Notes on Brain, Imagination and Creativity." In K. H. Pfenninger and V. R. Shubik (eds.), *The Origins of Creativity*, 59–68. Oxford: Oxford University Press.

Damerow, P. (1988). "Individual Development and Cultural Evolution of Arithmetical Thinking. In S. Strauss (ed.), *Ontogeny, Phylogeny, and Historical Development*. Norwood: Ablex, 125–152.

Damerow, P. (1998). "Prehistory and Cognitive Development." In J. Langer and M. Killen (eds.), *Piaget, Evolution, and Development*. Mahwah: Erlbaum, 247-269.

Dennett, D. (1969). *Content and Consciousness*. London: Routledge & Kegan Paul.

Derrida, J. (1962/1989). *Edmund Husserl's "Origin of Geometry": An Introduction*. London: University of Nebraska Press.

Dewey, J. (1916). *Essays in Experimental Logic*. New York: Dover Publications, INC.

Dewey, J. (1925/1929). *Experience and Nature*. New York: Norton and Company, Inc.

Dewey, J. (1934/2008). *Art as Experience*. Carbondale, IL: Southern Illinois University Press.

Dewey, J. (1938). *Experience and Education*. New York: Macmillan Company.

Dietrich, A. (2004). "The Cognitive Neuroscience of Creativity." *Psychonomic Bulletin & Review*, 11(6), 1001–26.

Di Paolo, E. A. (2018). "The Enactive Conception of Life." In A. Newen, S. Gallagher, and L. de Bruin (eds.), *The Oxford Handbook of Cognition: Embodied, Embedded, Enactive and Extended*, 71–94. Oxford: Oxford University Press.

Di Paolo, E. A. (2021). "Enactive Becoming." *Phenomenology and the Cognitive Sciences*, 20, 783–809.

Di Paolo E. A. and De Jaegher, H. (2022). "Enactive Ethics: Difference Becoming Participation." *Topoi*, 41, 241–256.

Di Paolo, E. A., Buhrmann, T., and Barandiaran, X. E. (2017). *Sensorimotor Life: An Enactive Proposal*. Oxford: Oxford University Press

Di Paolo, E. A., Cuffari, E. C., and De Jaegher, H. (2018). *Linguistic Bodies: The Continuity between Life and Language*. Cambrige (MA)/London: The MIT Press.

Dokic, J. and Arcangeli, M. (2015). "The Heterogeneity of Experiential Imagination." In T. Metzinger and J. M. Windt (eds.), *Open MIND*: 11(T). Frankfurt am Main: MIND Group.

Donald, M. (1991). *Origins of the Modern Mind*. Cambridge, MA: Harvard University Press.

Duncan, J. and Owen, A. M. (2000). "Common Regions of the Human Frontal Lobe Recruited by Diverse Cognitive Demands." *Trends in Neurosciences*, 23, 475–83.

Edelman, G. (1994). *Bright Air, Brillant Fire - On the Matter of the Mind*. Harmondsworth: Penguin.

Ehrson, H. H., Geyer, S., and Naito, E. (2003). "Imagery of Voluntary Movement of Fingers, Toes, and Tongue Activates Corresponding Body-Part-Specific Motor Representations." *Journal of Neurophysiology*, 90(5), 3304–16.

Ekstein, M. (2001). *Visions of a Compassionate World: Guided Imagery for Spiritual Growth and Social Transformation*. New York: Urim Publications.

Ellul, J. (1954). *La Technique ou l'enjeu du siècle* (The Technological Society). Paris: Colin.

Ellul, J. (1977). *Le Système technicien* (The Technological System). Paris: Calman-Lévy.

Espinas, A. (1890). "Les origines de la technologie" (The Origins of Technology). *Revue Philosophique de la France Et de l'Étranger*, 30, 113.

Fabre, M. (2005). "Deux sources de l'épistémologie des problèmes: Dewey et Bachelard" (Two Sources of the Epistemology of Problems: Dewey and Bachelard). *Les Sciences de l'éducation - pour l'ère nouvelle*, 3(38), 53–67.

Fauconnier, G. and Turner, M. B. (1998). "Conceptual Integration Networks." *Cognitive Science*, 22(2), 133–87.

Felleman, D. J., Van Essen, D. C. (1991). "Distributed Hierarchical Processing in the Primate Cerebral Cortex." *Cerebral Cortex*, 1(1), 1–47. https://doi.org/10.1093/cercor/1.1.1-a.

Feltz, D. and Landers, D. M. (2007). "The Effects of Mental Practice on Motor Skill Learning and Performance: A Meta-analysis." *Journal of Sport Psychology*, 5(1), 25–57.

Feltz, D. L. and Landers, D. M. (1983). "The Effects of Mental Practice on Motor Skill Learning and Performance: A Meta-Analysis." Journal of Sport Psychology, 5(1), 25–57.

Finke, R. A. (1986), "Mental Imagery and the Visual System." *Scientific American*, 254, 88–95.

Fodor, J. A. (1975). *The Language of Thought*. New York: Thomas Crowell.

Fodor, J. A. and Pylyshyn, Z. (1981). "How Direct is Visual Perception? Some Reflections on Gibson's 'Ecological Approach.'" *Cognition*, 9, 207–46.

Foglia, L. and Grush, R. (2011). "The Limitations of a Purely Enactive (Non-representational) Account of Imagery." *Journal of Consciousness Studies*, 18(5–6), 35–43.

Fox, P. T., Mintun, M. A., Raichle, M. E., Miezin, F. M., Allman, J. M., and Van Essen, D. C. (1986). "Mapping Human Visual Cortex with Positron Emission Tomography." *Nature*, 323(6091), 806–9.

Froese T. and Stewart J. (2012). "Enactive Cognitive Science and Biology of Cognition: A Response to Humberto Maturana." *Cybernetics & Human Knowing*, 19(4), 61–74

Gadamer, H.-G. (1960/2004). *Truth and Method*. London/New York: Continuum.

Galimberti, U. (1999). *Psiche e techne. L'uomo nell'età della tecnica* (Psyche and Techne. Man in the Age of Technology). Milan: Feltrinelli.

Gallagher, S. (2012). "Multiple Aspects in the Sense of Agency." *New Ideas in Psychology*, 30(1), 15–31.

Gallagher, S. (2016). "Pragmatic Interventions into Enactive and Extended Conceptions of Cognition." In R. Madzia and M. Jung (eds.), *Pragmatism and Embodied Cognitive Science*. Berlin/boston: De Gruyter.

Gallagher, S. (2017). *Enactivist Interventions: Rethinking the Mind*. Oxford: Oxford University Press.

Gallagher, S. (2020). *Action and Interaction*. Oxford: Oxford University Press.

Gallagher, S. and Bower, M. (2014). "Making Enactivism More Embodied." *Avant*, 5(2), 232–47.

Gallagher, S. and Crisafi, A. (2009). "Mental Institutions." *Topoi*, 28, 45–51.

Gallagher, S. and Rucińska, Z. (2020). "Making Imagination Even More Embodied: Imagination, Constraint and Epistemic Relevance." *Synthese*, 199, 8143–70.

Gallagher, S. and Rucińska, Z. (2021). "Prospecting Performance: Rehearsal and the Nature of Imagination." *Synthese*, 199, 4523–41.

Gallagher, S. and Zahavi, D. (2020). *The Phenomenological Mind*, 3rd edn. New York: Routledge.

Gallese, V., and Lakoff, G. (2005). "The Brain's Concepts: The Role of the Sensory-Motor System in Reason and Language." *Cognitive Neuropsychology*, 22, 455–79.

Gallina, J.-M. (2006). *Les représentations mentales. Une introduction accessible sur la notion de représentation mentale telle qu'elle a été introduite en psychologie cognitive, et telle qu'elle continue de soulever de nombreux débats* (Mental Representations. An Accessible Introduction to the Notion of Mental Representation as it has been Introduced in Cognitive Psychology, and as it Continues to Raise Many Debates). Éditions Dunod.

García, E. and Arandia, I. R. (2022). "Enactive and Simondonian Reflections on Mental Disorders." *Frontiers in Psychology*, 13, 938105.

Gardner, H. (1985). *The Mind's New Science: A History of the Cognitive Revolution*. New York: Basic Books.

Gaston Bachelard, La Formation de l'esprit scientifique. Contribution à une psychanalyse de la connaissance objective, Paris, Vrin, 1938.

Gaut, B. (2003). "Creativity and Imagination." In B. Gaut and P. Livingston (eds.), *The Creation of Art*, 268–93. Cambridge: Cambridge University Press.

Gaut, B. (2007). *Art, Emotion and Ethics*. Oxford: Oxford University Press.

Gaut B. (2010). " The Philosophy of Creativity." *Philosophy Compass*, 5(12), 1034–46. https://doi.org/10.1111/j.1747-9991.2010.00351.x.

Gaut, B. (2016). "Cinematic Art and Technology." In K. Thomson-Jones (ed.), *Current Controversies in Philosophy of Film*, 17–35. London: Routledge Taylor and Francis Group.

Gibson, J. J. (1979). *The Ecological Approach to Visual Perception*. Boston, MA: Houghton Mifflin.

Godoy, R. I. and Jørgensen, H., eds. (2001). *Musical Imagery*. Lisse: Swetz and Zeitlinger.

Goldman, A. (2006a). *Simulating Minds: The Philosophy, Psychology, and Neuroscience of Mindreading*. Oxford: Oxford University Press.

Goldman, A. (2006b). "Imagination and Simulation in Audience Responses to Fiction." In S. Nichols (ed.), *The Architecture of the Imagination: New Essays on Pretense, Possibility, and Fiction*, 41–56. Oxford: Oxford University Press.

Goldstein, K. (1934/1995). *The Organism: A Holistic Approach to Biology Derived From Pathological Data in Man*. New York: Zone Books.

Goodman, N. and Elgin, C. Z. (1988). *Reconceptions in Philosophy and Other Arts and Sciences*. Indianapolis: Hackett Publishing.

Gordon, R. M. (1995). "Folk Psychology as Simulation." In M. Davies and T. Stone (eds.), *Folk Psychology*, 60–73. Oxford: Blackwell.

Gottlieb, G. (2002). "Probabilistic Epigenesis of Development." In J. Valsiner and K. J. Connolly (eds.), *Handbook of Developmental Psychology*, 3–17, London: Sage.

Gottlieb, G. (2003). "On Making Behavioral Genetics Truly Developmental." *Human Development*, 46, 337–55.

Gottlieb, G. (2007). "Probabilistic Epigenesis." *Developmental Science*, 10(1), 1–11.

Guillot, A. and Collet, C. (2008). "Construction of the Motor Imagery Integrative Model in Sport: A Review and Theoretical Investigation of Motor Imagery Use." *International Review of Sport and Exercise Psychology*, 1(1), 31–44.

Guillot, A., Hoyek, N., Louis, M., and Collet, C. (2012). "Understanding the Timing of Motor Imagery: Recent Findings and Future Directions." *International Review of Sport and Exercise Psychology*, 5(1), 3–22.

Halpern, A. R., Zatorre, R., Bouffard, M., Johnson, J. A. (2004). "Behavioral and Neural Correlates of Perceived and Imagined Musical Timbre." *Neuropsychologia*, 42(9), 1281–92.

Hannay, A. (1971). "Mental Images, a Defence." *Revue Philosophique de la France Et de l'Etranger*, 162, 463–64.

Haugeland, J. (1990). "The Intentionality All-Stars." *Philosophical Perspectives*, 4, 383–427.

Haugeland, J. (1991). "Representation Genera." In W. Ramsey, D. E. Rumelhart, and S. Stich (eds.), *Philosophy and Connectionist Theory*, 33–59. Hillsdale, NJ: Lawrence Erlbaum.

Haugeland, J. (1998). *Having Thought: Essays in the Metaphysics of Mind*. Cambridge, MA: Harvard University Press.

Havelange, V. (2010). "The Ontological Constitution of Cognition and the Epistemological Constitution of Cognitive Science: Phenomenology, and Technology." In J. Stewart, O. Gapenne and E. A. Di Paolo (eds.), *Enaction: Toward a New Paradigm for Cognitive Science*, 335–9. Cambridge, MA: MIT Press.

Havelange V., Lenay C. and Stewart J. (2003). "Les représentations : mémoire externe et objets techniques." *Intellectica. Revue de l'Association pour la Recherche Cognitive*, 35(2), 115–219. Représentations: Quelques aperçus/ Longo : Laplace, Turing et le "jeu de l'imitation".

Heeger, D. J., Boynton, G. M., Demb, J. B., Seidemann, E., and Newsome, W. T. (1999). "Motion Opponency in Visual Cortex." *Journal of Neuroscience*, 19(16), 7162–74.

Hinton, G. (1979). "Some Demonstrations of the Effects of Structural Descriptions in Mental Imagery." *Cognitive Science*, 3, 231–50. https://doi.org/10.1207/s15516709cog0303_3.

Holmes, P. (2005). "Imagination in Practice: A Study of the Integrated Roles of Interpretation, Imagery and Technique in the Learning and Memorization Processes of Two Experienced Solo Instrumentalists." Unpublished MA Dissertation: University of Sheffield, UK.

Hopkins, L., Brookes, F., and Green, J. (2013). "Books, Bytes and Brains: The Implications of New Knowledge for Children's Early Literacy Learning." *Austrian Journal Early Childhood*, 38, 23–28.

Horrowitz, M. J. (1970a). *Image Formation and Cognition*. New York: Appleton-Century Crofts.

Horowitz, M. J. (1970b). *Subjective Phenomena: Image Formation and Cognition*. New York: Appleton-Century-Crofts.

Horowitz, M. J. (1983). *Image Formation and Psychotherapy*. New York: Jason Aronson.

Hurley, S. (1998). *Consciousness in Action*. Cambridge, MA: Harvard University Press.

Husserl, *Origin of Geometry*, in Derrida, J. (1962/1989). *Edmund Husserl's "Origin of Geometry": An Introduction*. London: University of Nebraska Press.

Hutchins, E. (1995a). *Cognition in the Wild*. Cambridge: MIT Press.

Hutchins, E. (1995b). "How a Cockpit Remembers its Speeds." *Cognitive Science*, 19, 265–88.

Hutchins, E. (2005). "Material Anchors for Conceptual Blends." *Journal of Pragmatics*, 37, 1555–77.

Hutchins, E. (2010). "Cognitive Ecology." *Topics in Cognitive Science*, 2(4), 705–15.

Hutchins, E. (2014). "The Cultural Ecosystem of Human Cognition." *Philosophical Psychology*, 27(1), 34–49.

Hutto, D. D. (2006). "Impossible Problems and Careful Expositions: Reply to Myin and De Nul." In R. Menary (ed.), *Radical Enactivism: Focus on the Philosophy of Daniel D. Hutto*, 25–64. Amsterdam: John Benjamins Publishing Company.

Hutto D. D. (2007). *Folk Psychological Narratives: The Sociocultural Basis of Understanding Reasons*. Cambridge, MA: MIT Press.

Hutto, D. D. (2015). "Overly Enactive Imagination? Radically Re-Imagining Imagining." *Southern Journal of Philosophy*, 53(S1), 68–89.

Hutto, D. D. and Myin, E. (2017). *Evolving Enactivism: Basic Minds Meet Content*. Cambridge, MA: MIT Press.

Hutto, D. D. and Satne, G. L. (2015). "The Natural Origins of Content." *Philosophia*, 43(3), 521–36.

Ingold, T. (2013a). *Making, Anthropology, Archaeology, Art and Architecture*. Londres: Routledge.

Ingold, T. (2013b). "Dreaming of Dragons." *Journal of the Royal Anthropological Institute*, 19, 734–52. https://doi.org/10.1111/1467-9655.12062.

Ingold, T. (2014). "The Creativity of Undergoing." *Pragmatics and Cognition*, 22(1), 124–39.

Jacobson, E. (1930). "Electrical Measurements of Neuromuscular States During Mental Activities. I. Imagination of Movement Involving Skeletal Muscle." *American Journal of Physiology*, 91, 567–608.

Jacobson, E. (1932). "Electrophysiology of Mental Activities." *The American Journal of Psychology*, 44, 677–94.

Jeannerod, M. (1997). *The Cognitive Neuroscience of Action*. Cambridge, MA: Blackwell Publishing.

Jennings, H. (1906). *Behavior of the Lower Organisms*. New York: Columbia University Press.

Johnson, M. (1987). *The Body in the Mind: The Bodily Basis of Meaning, Imagination, and Reason*. Chicago: University of Chicago Press.

Kant, I. (1790/2007). *Critique of Judgement*. Oxford: Oxford University Press.

Kaplan, D. (1968). "Quantifying in." *Synthese*, 19(1/2), 178–214.

Kaptelinin, V. and Nardi, B. A. (2006). *Acting with Technology: Activity Theory and Interaction Design*. Cambridge, MA: The MIT Press.

Käufer, S. and Chemero, A. (2015). *Phenomenology: An Introduction*. London: Polity Press.

Kelly, C. and Garavan, H. (2005). "Human Functional Neuroimaging of Brain Changes Associated with Practice." *Cerebral Cortex*, 15(8), 1089–102.

Kind, A. (2001). "Putting the Image Back in Imagination." *Philosophy and Phenomenological Research*, 62(1), 85–109.

Kind, A., ed. (2016). *Routledge Handbook of Philosophy of Imagination*. New York: Routledge.

Kind, A. and Kung, P., eds. (2016). *Knowledge Through imagination*. Oxford: Oxford University Press.

Kirsh, D. (1999). "L'utilisation intelligente de l'espace" (The Intelligent Use of Space). In M. de Fornel and L. Quéré (dirs.), *La logique des situations. Nouveaux regards sur l'écologie des activités sociales. Raisons Pratiques* 10, 227–60. Editions de l'EHESS.

Kirsh, D. (2005). "Metacognition, Distributed Cognition and Visual Design." In P. Gärdenfors and P. Johansson (eds.), *Cognition, Education, and Communication Technology*, 147–79. London: Routledge.

Kitcher, P. (1981). « Explanatory Unification ». In Philosophy of Science, 48(4), 507-31, by the Philosophy of Science Association.

Kiverstein, J. D. and Rietveld, E. (2018). "Reconceiving Representation-Hungry Cognition: An Ecological-enactive Proposal." *Adaptive Behavior*, 26(4), 147–63.

Köhler, W. (1917/1976). *The Mentality of Apes*. New York: Liveright.

Köhler, W. (1929). *Gestalt Psychology*. New York: Liveright.

Korn, E. R. and Johnson, K. (1983). *Visualization: The Uses of Imagery in the Health Professions*. Homewood, Ill: Dow Jones-Irwin.

Kosslyn, S. M. (1980). *Image and Mind*. Cambridge, MA: Harvard University Press.

Kosslyn, S. M. (1994). *Image and Brain: The Resolution of the Imagery Debate*. Cambridge, MA: MIT Press.

Kosslyn, S. M. and Ochsner, K. N. (1994). "In Search of Occipital Activation during Visual Mental Imagery." *Trends in Neurosciences*, 17(7), 290–2.

Kosslyn, S. M. and Pomerantz, J. (1977). "Imagery, Propositions and the Form of Internal Representations." *Cognitive Psychology*, 9, 52–76.

Kosslyn, S. M., Ball, T. M., and Reiser, B. J. (1978). "Visual Images Preserve Metric Spatial Information: Evidence From Studies of Image Scanning." *Journal of Experimental Psychology: Human Perception and Performance*, 4(1), 47–60.

Kosslyn, S. M., Reiser, B. J., Farah, M. J., and Fliegel, S. L. (1983). "Generating Visual Images: Units and Relations." *Journal of Experimental Psychology: General*, 112(2), 278–303. https://doi.org/10.1037/0096-3445.112.2.278.

Kosslyn, S. M., Ganis, G., and Thompson, W. (2001). "Neural Foundations of Imagery." *Neuroscience*, 2, 635–42.

Kosslyn, S. M., Thompson, W. L., and Ganis, G. (2006). *The Case for Mental Imagery*. Oxford: Oxford University Press.

Koukouti, M. D. and Malafouris, L. (2020a). "Material Imagination: An Anthropological Perspective." In A. Abraham (ed.), *The Cambridge Handbook of the Imagination*, 30–46. Cambridge: Cambridge University Press.

Koukouti, M. D. and Malafouris, L. (2020b). *An Anthropological Guide to the Art and Philosophy of Mirror Gazing*. London: Bloomsbury Publishing.

Kupiec, J.-J. (2009). *The Origin of Individuals: A Darwinian Approach to Developmental Biology*. Singapore: World Scientific Publishing.

Lakoff, G. and Johnson, M. (1999). *Philosophy in the Fltesh: The Embodied Mind and its Challenge to Western Thought*. New York: Basic Books.

Langland-Hassan, P. (2015). "Imaginative Attitudes." *Philosophy and Phenomenological Research*, *90*(3), 664–86.

Langland-Hassan, P. (2020). *Explaining Imagination*. Oxford: Oxford University Press.

Lenay, C. (2006). "Énaction, externalisme et suppléance perceptive." *Intellectica*, 43, 27–52.

Lenay, C. (2008). "Médiations techniques des interactions perceptives: Rencontres tactiles dans les environnements numériques partagés" (Technical Mediations of Perceptive Interactions: Tactile Encounters in Shared Numeric Environments). *Social Science Information*, 47(3), 331–52.

Leroi-Gourhan, A. (1964/1993). *Le Geste et la parole* (Gesture and Speech). Cambirge (MA)/London: The MIT Press.

Loh, K. K. and Kanai, R. (2016). "How has the Internet Reshaped Human Cognition?" *Neuroscientist*, 22, 506–20.

Lotze, M. and Halsband, U. (2006). "Motor Imagery." *Journal of Physiology-Paris*, 99(4–6): 386–95.

Lotze, M., Scheler, G., Tan, H.R., Braun, C., and Birbaumer, N. (2003). "The Musician's Brain: Functional Imaging of Amateurs and Professionals during Performance and Imagery." *Neuroimage*, 20, 1817–29.

Macherey, P. (1994). *Introduction à l'Éthique de Spinoza* (Introduction to Spinoza's Ethics). Paris: Presses Universitaires de France.

Maestri, E. and Antoniadis, P. (2015). "Notation as Instrument: From Representation to Enaction." *TENOR Proceedings International Conference 2015*, Paris-Sorbonne et IRCAM.

Malafouris, L. (2007). "Before and Beyond Representation: Towards an Enactive Conception of the Paleolithic Image." In C. Renfrew and I. Morley (eds.), *Image and Imagination: A Global Prehistory of Figurative Representation* 289–302. Cambridge: McDonald Institute for Archaeological Research.

Malafouris, L. (2013). *How Things Shape the Mind: A Theory of Material Engagement*. Cambridge: MIT Press.

Malafouris, L. (2014). "Creative Thinging: The Feeling of and for Clay." *Pragmatics and Cognition*, 22(1), 140–58.

Malafouris, L. (2015). "Metaplasticity and the Primacy of Material Engagement." *Time & Mind*, 8(4), 351–71.

Malafouris, L. (2018). "Bringing Things to Mind: 4Es and Material Engagement." In Albert Newen, Leon de Bruin, and Gallagher Shaun (eds.), *The Oxford Handbook of 4E Cognition*, 755–71. Oxford: Oxford University Press.

Malafouris, L. (2019a). "Mind and Material Engagement." *Phenomenology and the Cognitive Sciences*, 18(1), 1–17.

Malafouris, L. (2019b). "What Does the Stick do for the Blind?" In Jill Bennett and Mary Zournazi (eds.), *Thinking in the World*, 115–28. London: Bloomsbury Academic.

Malafouris, L. (2020). "Thinking as 'Thinging': Psychology with Things." *Current Directions in Psychological Science*, 29(1), 3–8. https://doi.org/10.1177/0963721419873349

Malafouris, L. (2021a). "How does Thinking Relate to Tool Making?" *Adaptive Behavior*, 29, 107–21. https://doi.org/10.1177/1059712320950539

Malafouris, L. (2021b). "Mark Making and Human Becoming." *Journal of Archaeological Method and Theory*, 28, 95–119. https://doi.org/10.1007/s10816-020-09504-4

Malafouris, L. (2021c). "Making Hands and Tools: Steps to a Process Archaeology of Mind." *World Archaeology*, 1–18.

Markman, K. D., Klein, W. M. P., and Suhr, J. A., eds. (2009). *Handbook of Imagination and Mental Simulation*. New York: Psychology Press.

Marr, D. (1980). *Vision, a Computational Investigation into the Human Representation and Processing of Visual Information*. San Francisco: Freeman.

Martin, M. (2002). "The Transparency of Experience." *Mind and Language*, 17(4), 376–425.

Martinez S. F. and Villanueva, A. (2018). "Musicality as Material Culture." *Adaptive Behavior*, 26(5), 257–67.

Maturana, H. (2002). "Autopoiesis, Structural Coupling and Cognition: A History of These and Other Notions in the Biology of Cognition." *Cybernetics & Human Knowing*, 9(3–4), 5–34.

Maturana, H. R., and Varela, F. J. (1972). *Autopoiesis and Cognition: The Realization of the Living*. Dordrecht, Holland: D. Reidel Publishing.

Maturana, H. and Varela, F. J. (1980). *Autopoiesis and Cognition: The Realization of the Living*. Dordrecht, Holland: D. Reidel Publishing.

Maturana, H. and Varela, F. (1987). *The Tree of Knowledge. The Biological Roots of Human Understanding*. Boston and London: Shambhala.

McLean, S. (2009). "Stories and Cosmogonies: Imagining Creativity beyond 'Nature' and 'Culture'".*Cultural Anthropology*, 24(2), 213–45. https://doi.org/10.1111/j.1548-1360.2009.01130.x.

McGinn, C. (2004). *Mindsight: Image, Dream, Meaning*. Cambridge, MA: Harvard University Press.

Medina, J. (2013). "An Enactivist Approach to the Imagination: Embodied Enactments and 'Fictional Emotions'." *American Philosophical Quarterly*, 50(3), 317–35.

Mellet, E., Petit, L., Mazoyer, B., Denis, M., and Tzourio, N. (1998). "Reopening the Mental Imagery Debate: Lessons from Functional Anatomy." *Neuro-Image*, 8, 129–39.

Merleau-Ponty, M. (1942/1963). *The Structure of Behavior*. Boston: Beacon Press.
Merleau-Ponty, M., (1945/2012). *Phenomenology of Perception*. London: Routledge.
Merleau-Ponty, M., (1957–1958/1995). *Nature and Logos: Course Notes from the College de France*. Evanston: Northwestern University Press.
Merleau-Ponty, M. (1964). *Le Visible et l'invisible*, publié par Cl. Lefort, Paris: Gallimard.
Meshi, D., Tamir, D. I., and Heekeren, H. R. (2015). "The Emerging Neuroscience of Social Media." *Trends in Cognitive Science*, 19, 771–82.
Mills, K. L. (2014). "Effects of Internet Use on the Adolescent Brain: Despite Popular Claims, Experimental Evidence Remains Scarce." *Trends in Cognitive Science*, 18, 385–7.
Modrak, D. K. W. (2016). "Aristotle on Phantasia." In A. Kind (ed.), *Routledge Handbook of Philosophy of Imagination*, 15–26. London: Routeldge.
Moran, A. (1994). "The Expression of Feeling in Imagination." *The Philosophical Review*, 103(1).
Moran, A. and O'Shea, H. (2020). "Motor Imagery Practice and Cognitive Processes." *Frontiers in Psychology*, 11, 394.
Moyal-Sharrock, D. (2009). "The Fiction of Paradox: Really Feeling for Anna Karenina." In Y. Gustafsson, C. Kronquist, and M. McEachrane (eds.), *Emotions & Understanding: An Anthology*, 165–84. London: Palgrave Macmillan.
Munzert, J. Lorey, B., and Zentgraf, K. (2009). "Cognitive Motor Processes: The Role of Motor Imagery in the Study of Motor Representations." *Brain Research Review*, 60(2), 306–26.
Mulligan, K. (1999). "La varietà e l'unità dell'immaginazione" (The Variety and Unity of the Imagination). *Rivista di Estetica*, 11(2), 53–67.
Myin, E. and O'Regan, J. K. (2002). "Perceptual Consciousness, Access to Modality and Skill Theories: A Way to Naturalize Pphenomenology?" *Journal of Consciousness Studies*, 9(1), 27–45.
Nagel, T. (1974). "What it is Like to be a Bat?" *Philosophical Review*, 83(4), 435–50.
Neisser, U. (1976). *Cognition and Reality: Principles and Implications of Cognitive Psychology*. New York: WH Freeman and Company.
Neisser, U. and Kerr, N. (1973). "Spatial and Mnemonic Properties of Visual Images." *Cognitive Psychology*, 5(2), 138–50.
Nichols, S. (2006). "Imaginative Blocks and Impossibility: An Essay in Modal Psyschology." In Sh. Nichols (ed.), *The Architecture of the Imagination*, 237–55. Oxford: Oxford University Press.
Nichols, S. (2009). "The Propositional Imagination." In *The Routledge Companion to Philosophy of Psychology*, John Symons and Paco Calvo, 360–9. Routledge, LONDON AND NEW YORK .
Nichols, S. and Stich, S. P. (2003). *Mindreading: An Integrated Account of Pretense, Self-awareness and Understanding Other Minds*. Oxford: Oxford University Press.
Nicolas of Cusa. (1453/2007). *The Vision of God*, trans. E. G. Salter. New York: Cosimo.

Nietzsche, F. (1878/1996). *Human, All Too Human*, trans. R. J. Hollingdale, Cambrige: Cambrige University Press.

Noordhof, P. (2002). "Imagining Objects and Imagining Experiences." *Mind and Language*, 17(4), 426–55.

Nyberg, L. (1998). "Mapping Episodic Memory." *Behavioural Brain Research*, 90, 107–14.

O'Brien, J. and Opie, J. (2008). "The Role of Representation in Computation." *Cognitive Processing*, 10, 53–62.

O'Craven, K. M. and Kanwisher, N. (2000). "Mental Imagery of Faces and Places Activates Corresponding Stimulus-Specific Brain Regions." *Journal of Cognitive Neuroscience*, 12(6), 1013–23.

Opie, J. (2008). "The Role of Representation in Computation." *Cognitive Processing*, 10, 53–62.

O'Regan, J. K. (1992). "Solving the 'Real' Mysteries of Visual Perception: The World as an outside Memory." *Canadian Journal of Psychology*, 46(3), 461–88.

O'Regan, J. K. and Noë, A. (2001). "A Sensorimotor Approach to Vision and Visual Perception." *Behavioral and Brain Sciences*, 24(5), 939–73.

O'Shaughnessy, B. (1980). *The Will: A Dual Aspect Theory*. Cambridge: Cambridge University Press.

Paivio, A. (1971). "Imagery and Language." In J. S. Segal (ed.), *Imagery: Current Cognitive Approaches*. Elsevier Inc,S. Segal (ed) Academic Press, New York.

Peacocke, C. (1985). "Imagination, Possibility and Experience: A Berkeleian View Defended." In J. Foster and H. Robinson (eds.), *Essays on Berkeley*, 19–35. Oxford: Oxford University Press.

Pfenninger, K. H. and Shubik, V. R. (2001). "Insights into the Foundation of Creativity: A Synthesis." In K. H. Pfenninger and V. R. Shubik (eds.), *The Origins of Creativity*, 213–36. Oxford: Oxford University Press.

Piaget, J. (1962). *Play, Dreams, and Imitation in Childhood*. New York Norton.

Pitt, D. (2020). "Mental Representation." In N. Edward (ed.), *The Sandford Encyclopedia of Philosophy* (Spring 2020). URL = https://plato.stanford.edu/archives/spr2020/entries/mental-representation/

Poldrack, R. A. (2000). "Imaging Brain Plasticity: Conceptual and Methodological Issues." *Neuroimage*, 12(1), 1–13.

Popper, K. (1934). *Logik der Forschung*. (The Logic of Scientific Discovery). Vienna: Julius Springer, Hutchinson & Co.

Poulsgaard, K. S. (2019). "Enactive Individuation: Technics, Temporality and Affect in Digital Design and Fabrication." *Phenomenology of the Cognitive Science*, 18(1), 281–98.

Pylyshyn, Z. W. (1973). "What the Mind's Eye Tells the Mind's Brain: A Critique of Mental Imagery." *Psychological Bulletin*, 80, 1–25.

Pylyshyn, Z. W. (1978). "What Has Language to Do with Perception? Some Speculations on the Lingua Mentis." In *Theoretical Issues in Natural Language Processing-2*, 172–9. New York: ACM.

Pylyshyn, Z. W. (1981). "The Imagery Debate: Analogue Media Versus Tacit Knowledge." *Psychological Review*, 88, 16–45.

Pylyshyn, Z. W. (1984). *Computation and Cognition*. Cambridge, MA: The MIT Press.

Pylyshyn, Z. W. (2002). "Mental Imagery: In Search of a Theory." Behavioral and Brain Sciences, 25(2), 157–238. https://doi.org/10.1017/S0140525X02000043.

Pylyshyn, Z. (2003a). "Return of the Mental Image: Are There Really Pictures in the Brain?" Trends in Cognitive Sciences, 7(3), 113–8. https://doi.org/10.1016/S1364-6613(03)00003-2.

Pylyshyn, Z. (2003b). *Seeing and Visualizing: It's Not What You Think*, 563. Cambridge, MA: MIT Press.

Ramsey, W. (2007). "Representation Reconsidered." *The British Journal of the Philosophy of Science*, 62(3), 669–75.

Ramsey, W. (2017). "Must Cognition be Representational?" *Synthese*, 194(11), 4197–214.

Richardson, J. (1999). *Imagery*. London: Psychology Press.

Robertson, T. (2006). "Ethical Issues in Interaction Design." *Ethics and Information Technology*, 8(2), 49–59.

Rideout, V. and Hamel, E. (2006). *The Media Family: Electronic Media in the Lives of Infants, Toddlers, Preschoolers, and Their Parents*. San Francisco, CA: Kaiser Family Foundation.

Rietveld, E. and Kiverstein, J. (2014). "A Rich Landscape of Affordances." *Ecological Psychology*, 26(4), 325–52.

Rousseau, J.-J. (1755). *Discours sur l'origine et les fondements de l'inégalité parmi les hommes* [Discourse on the Origin and Basis of Inequality Among Men].

Rowlands, M. (2003). *Externalism: Putting Mind and World Back together Again*. Chesham, England: Acumen.

Rucińska, Z. (2014). "Basic Pretending as Sensorimotor Engagement?" In J. M. Bishop and A. O. Martin (eds.), *Contemporary Sensorimotor Theory, Studies in Applied Philosophy, Epistemology and Rational Ethics*, 15, 175–87. New York: Springer.

Rucińska, Z. (2016). What Guides Pretense? Towards the Interactive and the Narrative Approaches." *Phenomenology and the Cognitive Sciences*, 15(1), 117–33.

Runco, M. (2010). "Parsimonious Creativity and Its Measurement." In E. Villaba (ed.), *Measuring Creativity: Proceedings of the European Council Meeting on Creativity and Innovation*. Luxembourg: Publications Office of the European Union.

Rupert, R. D. (2004). "Challenges to the Hypothesis of Extended Cognition." *Journal of Philosophy*, 101(8), 389–428.

Rupert, R. D. (2006). "Extended Cognition as a Framework for Empirical Psychology: The Costs Outweigh the Benefits." *American Philosophical Association Central Division Meeting*.

Rupert, R. D. (2008). "Causal Theories of Mental Content." *Philosophy Compass*, 3(2), 353–80.

Ryle, G. (1949/2009). *The Concept of Mind*. New York: Rootledge.

Sarbin, T. R., and Juhasz, J. B. (1970). "Toward a Theory of Imagination." *Journal of Personality*, 38(1), 52–76. https://doi.org/10.1111/j.1467-6494.1970.tb00637.x.

Sartre, J.-P. (1936). *The Imagination*. London and New York: Routledge.

Sartre, J.-P. (1940). *The Psychology of Imagination*, translated from the French by B. Frechtman, New York: Philosophical Library.

Sawyer, R. K. (2006). *Explaining Creativity: The Science of Human Innovation*. New York: Oxford University Press.

Schutz, A. (1951). "Making Music Together: A Study in Social Relationship." *Social Research*, 18, 76–97.

Sengpiel, F. and Hübener, M. (1999). "Spotlight on the Primary Visual Cortex." *Current Biology*, 9(9), R318–21.

Sereno, M. I., Dale, A. M. et al. (1995). "Borders of Multiple Human Visual Areas in Humans Revealed by Functional MRI." *Science*, 268, 889–93.

Serris, J.-P. (1994). *La technique* (Technics), Paris: Presses Universitaires de France.

Shoemaker, S. S. (1968). "Self-Reference and Self-Awareness." *Journal of Philosophy*, 65, 555–67.

Sheikh, A. A., Kunzendorf, R. G., Sheikh, K. S., and Baer, S. M. (2003). "Physiological Consequences of Imagery and Related Approaches." In A. A. Sheikh (ed.), *Healing Images: The Role of Imagination in Health*, 27–52. New York: Baywood Publishing Co.

Sills, H. (2005). "Time Remembered." *KronoScope*, 5, 59–72.

Simondon, G. (1954). Unpublished letter quoted in Nathalie Simondon's biography. http://gilbert.simondon.fr/content/biography.

Simondon, G. (1958a/2005). *L'Individuation à la lumière des notions de forme et d'information* (Individuation in the Light of the Notions of Form and Information). Grenoble: Jérôme Millon.

Simondon, G. (1958b/2012). *Du Mode d'existence des objets techniques* (Mode of Existence of Technical Objects). Lonrai: Aubier.

Simondon, G. (1961/2006), «La mentalité technique» (The Technical Mentality). *Sur la Technique*. Paris: Presses Universitaires de France.

Simondon, G. (1962/2010). « L'amplification dans les processus d'information » (Amplification in Information Processes). In Communication et information, Cours et conférences. Chatou: La Transparence,pp. 159-173.

Simondon, G. (1964–1965/2006). *Cours sur la perception* (Lessons on Perception). Chatou: La Transparence.

Simondon, G. (1965–1966/2008). *Imagination et invention* (Imagination and Invention). Chatou: La Transparence.

Simondon, G. (1968/2005). *L'invention dans les techniques* (Invention in Techniques), ed. J.-Y. Chateau. Paris: Le Seuil. https://www.seuil.com/ouvrage/l-invention-dans-les-techniques-gilbert-simondon/9782020563376.

Simondon, G. (1976/2005). *L'invention dans les techniques. Cours et conferences* (Invention in Techniques, Lessons and Conferences). Paris: Le Seuil.

Simonton, D. K. (2000). "Creativity: Cognitive, Personal, Developmental, and Social Aspects." *American Psychologist*, 55(1), 151–8.

Slezak, P. (1992). The "Philosophical" Case Against Visual Images. A "Crucial" Experiment, Center for Cognitive Science University of New South Wales, Kensington.

Slezak, P., Caelli, T., and Clark, R., eds. (1995). *Perspectives on Cognitive Science, Volume 1: Theories, Experiments, and Foundations*. New York: Ablex Publishing.

Smolensky, P. (1986). "The Proper Treatment of Connectionism." *Behavioral and Brain Science*, 11, 1–74.

Steiner, P. (2008a). "Délocaliser les phénomènes mentaux : la philosophie de l'esprit de Dewey" (Delocalizing Mental Phenomena: The Philosophy of Mind of Dewey). *Revue international de philosophie*, 3(245), 273–92.

Steiner, P. (2008b). "Sciences cognitives, tournant pragmatique et horizons pragmatistes" (Cognitive Sciences, Pragmatic Turn and Pragmatist Horizons). *Tracés*, 15, 85–105.

Steiner, P. (2010). "The Bounds of Representation: A Non-Representationalist us of the Resources of the Model of Extended Cognition." *Pragmatics and Cognition*, 18(2), 235–72.

Steiner, P. (2011). "Enaction, pragmatisme et minimalisme représentationnel" (Enaction, Pragmatism and Representational Minimalism). In N. Bault, V. Chambon, N. Maïonchi-Pino, F.-X. Pénicaud, B. Putois, J.-M. Roy (eds.), *Peut-on se passer de representations en sciences cognitives?* 191–210. Bruxelles-Paris: De Boeck.

Steiner, P. (2014). "The Delocalized Mind. Judgements, Vehicles, and Persons." *Phenomenology and the Cognitive Sciences*, 13(3), 437–60.

Steiner, P. (2016). "Embodied Cognitive Science, Pragmatism, and the Fate of Mental Representation." In R. Madzia and M. Jung (eds.), *Pragmatism and Embodied Cognitive Science*. Berlin/boston: De Gruyter.

Steiner, P. (2018), "Reading and Understanding: On Some Differences Between Wittgenstein and 4E Cognitive Science." *Rivista Italiana di Filosofia del Linguaggio*, 12(2), 124–37.

Steiner, P. (2019a). "De l'action à l'expression: Wittgenstein, les phénomènes mentaux et les sciences cognitives." *Archives de Philosophie*, 82, 361–82.

Steiner P. (2019b). *Désaturer l'Esprit. Usages du pragmatisme* (Desaturate the Mind. Uses of Pragmatism). Paris: Questions Théoriques.

Steiner, P. (2020). "Le débat entre représentationnalisme et antireprésentationnalisme cognitifs: de la métaphysique à la philosophie des sciences." In N. Depraz and R. Künstler (dir.) *Enquête sur les représentations mentales*. Éditions matériologiques.

Sternberg, R. J. and Lubart, T. I. (1999). "The Concept of Creativity: Prospects and Paradigms." In R. J. Sternberg (ed.), *Handbook of Creativity*, 3–15. Cambridge: Cambridge University Press.

Stevenson, L. (2003). "Twelve Conceptions of Imagination." *The British Journal of Aesthetics*, 43(3), 238–59.

Stewart, J. (1996). "Cognition = Life: Implications for Higher-Level Cognition." *Behavioural Processes*, 35, 311–26.

Stewart, J. (2001). "Radical Constructivism in Biology and Cognitive Science." *Foundations of Science*, 6(1–3), 99–124.

Stewart, J. (2010). "Foundational Issues in Enaction as a Paradigm for Cognitive Science: From the Origin of Life to Consciousness and Writing." In J. Stewart, O. Gapenne, and E. Di Paolo (eds.), *Enaction. Toward a New Paradigm for Cognitive Science*. Cambridge (MA): MIT Press.

Stewart, J. (2011). "Life as a Process of Bringing Forth a World." *Constructivist Foundations*, 7(1), 21–2.

Stewart, J. (2018). "Lived Experience: Past and Present." *Constructivist Foundations*, 13(2), 237–8.

Stiegler, B. (1994/1998). *Technics and Time, 1: The Fault of Epimetheus*. Stanford: Stanford University Press.

Stiegler, B. (1996/2009). *Technics and Time, 2: Disorientation*. Stanford: Stanford University Press.

Stiegler, B. (2001/2010). *Technics and Time, 3: Cinematic Time and the Question of Malaise*. Stanford: Sandford University Press.

Stiegler, B. (2005). *De la misère symbolique, II. La "catastrophe" du sensible*. Paris: Galilée.

Stock, K. (2017). *Only Imagine: Fiction, Interpretation, and Imagination*. Oxford: Oxford University Press.

Stokes, D. (2007). "Incubated Cognition and Creativity." *Journal of Consciousness Studies*, 14, 83–100.

Stokes, D. (2008). "A Metaphysics of Creativity." In K. Stock and K. Thomson Jones (eds.), *New Waves in Aesthetics*. Basingstoke: Palgrave Macmillan.

Stokes, D. (2011). "Minimally Creative Thought." *Metaphilosophy*, 42(5), 658–81.

Stokes, D. (2014). "The Role of Imagination in Creativity." In E. S. Paul and S. B. Kaufman (eds.), *The Philosophy of Creativity*. New York: Oxford University Press.

Stokes, D. (2016). "Imagination and Creativity." In A. Kind (ed.), *The Routledge Handbook of Philosophy of Imagination*, 247–61. Routledge/Taylor & Francis Group.

Strawson, P. F. (1970). "Imagination and Perception." In R. C. Sutherland Walker (ed.), *Kant on Pure Reason*. New York: Oxford University Press.

Theiler, A. M. and Lippman, L. G. (1995). "Effects of Mental Practice and Modeling on Guitar and Vocal Performance." *Journal of General Psychology*, 122(4), 329–43.

Thomas, N. J. T. (1999). "Are Theories of Imagery Theories of Imagination? An Active Perception Approach to Conscious Mental Content." *Cognitive Science*, 23(2), 207–45.

Thomas, N. J. T. (2010). "Mental Imagery." *The Stanford Encyclopedia of Philosophy*.

Thomas, N. J. T. (2014). "The Multidimensional Spectrum of Imagination: Images, Dreams, Hallucinations, and Active, Imaginative Perception." *Humanities*, 3(2), 132–84.

Thompson, E. (2007). *Mind in Life: Biology, Phenomenology and the Sciences of Mind.* Cambridge, MA: Harvard University Press.

Thompson, E. (2015). "The Brain Is Wider Than the Sky: What Emily Dickinson Can Teach Neuroscience." *Psychology Today*, April 16.

Tootell, R. B., Hadjikhani, N. K., Vanduffel, W., Liu, A. K. Mendola, J. D. Sereno, M. I., and Dale, A. M. (1998). "Functional Analysis of Primary Visual Cortex (V1) in Humans." *Proceedings of the National Academy of Sciences of the United States of America*, 95, 811–17.

Turner, Ph. (2020). *Imagination + Technology*. Cham: Springer.

Tusek, D., Church, J. M., and Fazio, V. W. (1997). "Guided Imagery as a Coping Strategy for Perioperative Patients." *AORN*, 66, 644–9.

Tye, M. (1991). *The Imagery Debate.* Cambridge, MA: The MIT Press.

Uncapher, M. R., Lin, L., Rosen, L. D., Kirkorian, H. L., Baron, N. S., Bailey, K., et al. (2017). "Media Multitasking and Cognitive, Psychological, Neural, and Learning Differences." *Pediatrics*, 140, 62–6.

Van Dijk, L. and Rietveld, E. (2018). "Situated Anticipation." *Synthese*, 198, 1–37.

Van Dijk, L. and Rietveld, E. (2020). "Situated Imagination." *The Phenomenology of the Cognitive Science*. https://doi.org/10.1007/s11097-020-09701-2

Van Dijk, L. and Withagen, R. (2016). "Temporalizing Agency: Moving Beyond on- and offline Cognition." *Theory and Psychology*, 26(1), 5–26.

Van Leeuwen, N. (2013). "The Meaning of "Imagine" Part I: Constructive Imagination." *Philosophy Compass*, 8(3), 220–30.

Van Leeuwen, N. (2016). In A. Kind (ed.), *The Routledge Handbook of Philosophy of Imagination*. New York: Routledge.

Van Rooij, I., Bongers, R., and Haselager, W. (2002). "A Non-Representational Approach to Imagined Action." *Cognitive Science*, 26(3), 345–75.

Varela, F. J. (1979). *Principles of Biological Autonomy*. New York, NY: North Holland.

Varela, F. J. (1991). "Organism: A Meshwork of Selfless Selves-." In A. I. Tauber (ed.), *Organism and the Origins of the Self*, 79–107. Dordrecht, The Netherlands: Kluwer.

Varela, F. J. (1997). "Patterns of Life: Intertwining Identity and Cognition." *Brain and Cognition*, 34, 72–87.

Varela, F. J. (1999). *Ethical Know-How: Action, Wisdom, and Cognition*. Stanford, CA: Stanford University Press.

Varela, F. J. (2000). *El Fenómeno de la Vida*. Santiago de Chile: J. C. Sáez.

Varela, F. J., Maturana, H. R., and Uribe, R. (1974). "Autopoiesis: The Organization of Living Systems, its Characterization and a Model." *BioSystems*, 5, 187–96.

Varela, F. J., Thompson, E., and Rosch, E. (1991). *The Embodied Mind: Cognitive Science and Human Experience*. Cambridge, MA: MIT Press.

Vedechkina, M. and Borgonovi, F. (2021). "A Review of Evidence on the Role of Digital Technology in Shaping Attention and Cognitive Control in Children." *Frontiers in Psychology*, 12, 611155.

Vendler, Z. (1984). *The Matter of Minds*. Oxford: Clarendon Press.

Viau, M. (2009). "La métaphore du miroir chez Nicolas de Cues" (The Mirror Metaphor in Nicolas of Cusa). *Revue des sciences religieuses*, 83(2), 257–76.

Von Eckardt, B. (1988). "Mental Images and Their Explanations." *Philosophical Studies: An International Journal for Philosophy in the Analytic Tradition*, 53(3), 441–60.

Von Eckardt, B. (1993). What Is Cognitive Science? Cambridge, MA: The MIT Press.

Vygotsky, L. (1934/1962). *Thinking and Speech*. Cambridge, MA: The MIT Press.

Walton, Kendall L. (1990). *Mimesis as Make-Believe: On the Foundations of the Representational Arts*. Cambridge, MA: Harvard University Press.

Walton, Kendall L. (2006). *Imagination: New Essays on Pretense, Possibility, and Fiction*. Oxford: Oxford University Press.

Ward, T. B., Smith, S. M., and Finke, R. A. (1999). "Creative Cognition." In R. J. Sternberg (ed.), *Handbook of Creativity*, 189–12. Cambridge: Cambridge University Press.

Weisberg, R. W. (1993). *Creativity: Beyond the Myth of Genius*. Brand: W.H. Freeman Co.

White, H. (1973). *Metahistory: The Historical Imagination in Nineteenth-Century Europe*. Baltimore: The Johns Hopkins University Press.

Wieman, H. N. (1961). *Intellectual Foundations of Faith*. London: Vision Press.

Willard, R. D. (1977). "Breast Enlargement through Visual Imagery and Hypnosis." *American Hypnosis*, 19, 195–200.

Williams, B. (1973). "Imagination and the Self." In *Problems of the Self*, 26–45. Cambridge: Cambridge University Press.

Wilson, R. (2004). *Boundaries of the Mind: The Individual in the Fragile Sciences*. Cambridge: Cambridge University Press.

Wittgenstein, L. (1953). *Philosophical Investigations*, trans. G. E. M. Anscombe, 3rd edn. Oxford: Blackwell.

Wollheim, R. (1984). *The Thread of Life*. Cambridge: Harvard University Press.

Young, J. Z. (1978). *Programs of the Brain*. Oxford: Oxford University Press.

Zatorre, R. -J. (1999). "Brain Imaging Studies of Musical Perception and Musical Imagery." In *Journal of New Music Research*, 28, 229–36.

Zatorre, R. J. and Halpern, A. R. (2005). "Mental Concerts: Musical Imagery and Auditory Cortex." *Neuron*, 47, 9–12.

Zatorre, R. J., Chen, J. L., and Penhune, V. B. (2007). "When the Brain Plays Music: Auditory-Motor Interactions in Music Perception and Production." *Nature Reviews Neuroscience*, 8(7), 547–58

Zatorre, R. J., Evans, A. C., and Meyer, E. (1994). "Neural Mechanisms Underlying Melodic Perception and Memory for Pitch." *Journal of Neuroscience*, 14, 1908–19.

Zimmermann-Shlatter, A., Schuster, C., Puhan, M. A., Siekierka, E., and Steurer, J. (2008). "Efficacy of Motor Imagery in Post-Stroke Rehabilitation: A Systematic Review." *Journal of NeuroEngineering and Rehabilitation*, 5, 8–10.

Index

www.ingramcontent.com/pod-product-compliance
Ingram Content Group UK Ltd.
Pitfield, Milton Keynes, MK11 3LW, UK
UKHW020812290926
15470UKWH00003B/1112

9 781350 507593